IMAGES of Aviation

STINSON AIRCRAFT COMPANY

STINSON
AIRCRAFT
"The Aircraft Standard of the World"
MANUFACTURED BY THE
STINSON AIRCRAFT CORP.
DETROIT, MICH.
U.S.A.

The Stinson logo, used in various forms during the life of the company, incorporated the line, "The Aircraft Standard of the World," a positioning statement borrowed from the Cadillac Motor Car Co. At that time, the middle 1920s, the Cadillac brand denoted the pinnacle of automobile performance, style, and luxury that the Stinson organization attempted to emulate in the airplanes it built and marketed.

After its founding the Stinson organization went through a number of corporate associations; going from the Stinson Syndicate through Corporation to a Division of Cord Corp., to Avco and Vultee to Convair; then dying nearly 25 years later under Piper's control. However, generations of Detroiters, having been close to the man and his organization, simply called it the Stinson Airplane Company.

IMAGES
of Aviation

STINSON AIRCRAFT COMPANY

John A. Bluth

ARCADIA
PUBLISHING

ISBN 978-0-7385-2020-9

Published by Arcadia Publishing
Charleston, South Carolina

Printed in the United States of America

Library of Congress Catalog Card Number: 2002112091

For all general information contact Arcadia Publishing at:
Telephone 843-853-2070
Fax 843-853-0044
E-mail sales@arcadiapublishing.com
For customer service and orders:
Toll-Free 1-888-313-2665

Visit us on the Internet at www.arcadiapublishing.com

Pictured here are two single engine Stinsons, each at the head of its class for its time. In the foreground is the post-World War II Model 108 Voyager powered by the flat, 6-cylinder, 150-hp Franklin engine. In the background is a classic Stinson 1936 Gullwing powered by a 350-hp Wright radial engine.

Contents

Acknowledgments

The gathering of information for this work began long before I had any firm idea of what I was going to do with it. Over the years, I came to realize that I had some ideas, information, and images that had not yet been revealed. In the spirit of wanting to share what I have gathered, this work is presented. Listed below are some of the people whose contributions breathed life into this project.

John Underwood, creator of the definitive book on the Stinson family; W.A. "Bill" Mara Jr., who shared family photos and insights; Virginia A. Lundquist, whose eagle eye scanned these pages; Tom Featherstone, Wayne State University, W. P. Reuther Library; Fran Gazlay and Sandy Basse, Northville Historical Society; Valerie Latzman, Wayne Historical Museum; Jon M. Bill, Auburn-Cord Duesenberg Museum; Sue Lurvey, EAA Library; Bud Manning, photographer; Tim O'Callaghan, author of *The Aviation Legacy of Henry and Edsel Ford*; Bob Taylor, founder of the Antique Airplane Association; Sydney Pemberton, Esq.; the Birmingham Library staff, with special mention of Sarah Ormond and Dolores Lowe, and the Technical Section staff of the Detroit Main Branch Library.

And, most importantly, Carol A. Bluth, whose corrections, suggestions, and direction have allowed this unguided missile to fly. Thanks.

Photo acknowledgements as follows:

John Underwood, pages 8, 9, 14, 58, 81, 100, 122, 123
Northville Public Library, pages 11, 46, 47, 72
W. A. Mara, Jr., pages 11, 21, 26
Robert Pauley, pages 13, 27, 31, 38, 59, 87, 102
Wayne Historical Society, pages 19, 68, 71, 88, 96, 121, 128
Tim O'Callaghan, pages 22, 24
Detroit Athletic Club, page 23
Burton Collection, Detroit Public Library, page 30
Auburn Cord Museum, pages 44, 67, 73, 75, 90, 101, 112, 114
Experimental Aircraft Association Library, pages 36, 82, 84, 97
Antique Airplane Association Library, pages 56, 57
Wayne State University, W. P. Reuther Library, pages 60, 79
Northville Historical Society, page 61
Bud Manning, page 127

FOREWORD

In 1925, *Aviation Magazine*, in its "Who's Who in American Aviation" section, described Stinson, Edward A., in part, in the following terms:

Commercial Aviation: born, Ft. Payne, Ala., July 11, 1894: Son of Edward A. Stinson and Emma B. (Beavers) Stinson: Married Estelle Judy Oct. 1, 1919.
Aeronautical Activities: Exhibition flying and instructing, 1912 to1917: Instructor, U.S. Army and test pilot, Curtiss Co., 1917 to 1919: Commercial flying, 1919 to 1925.
Flying Rating: F.A.I. Certificate. War Service, (World War I) Instructor, U.S. Army, 1916 to 1919.
Honors: World endurance record Mineola, L.I., N.Y., Dec. 30, 1921.
Present Occupation: Pres., Stinson Sales Corp.
Address: 8226 Wilson Ave., Detroit, Mich.

Edward A. Stinson Jr. was the second of four children born to Edward and Emma Stinson, all of whom became involved with aviation. Katherine, three years Eddie's senior, blazed the trail. Convincing her parents that flying lessons were more important than learning piano, she was instructed by famed aviator Max Lillie, based in Chicago. After four hours of dual instruction, she soloed and shortly thereafter, at age 21, was awarded a Federation Aeronautique International Certificate, becoming a "licensed" pilot on July 12, 1912. Katherine then began a series of highly profitable flying exhibitions around the country and established the "Stinson" name in the forefront of recognized flyers. The combination of comely Katherine with the thrill of seeing actual aircraft flight drew huge crowds wherever she performed.

Pictured here is the Stinson family of flyers: Katherine, Eddie, and Marjorie as they looked in about 1915. After staging more exhibitions during the summer, Katherine established a flying school in San Antonio, Texas, occupying some unused sheds on the grounds of Fort Sam Houston. Brothers Eddie, Jack B., and sister Marjorie soon joined her. Eddie contributed to the family effort by providing mechanical service on Katherine's balky flying machines. Eddie wanted to learn to fly but his sisters refused to teach him. Eddie had already gathered a reputation of being a diligent worker but also of being reckless and all too fond of cigarettes and spirits.

Undaunted, Eddie gathered his meager finances and went to Dayton, Ohio, to learn to fly at the Wright school. After some lessons, he returned to San Antonio to continue work on the school's planes and gather more flying hours. Elsewhere in 1914, World War I began in Europe with air power quickly becoming an important military tool. In the U.S. and Canada, there was little action taken to develop an air force. Canadians wanting to qualify for England's Royal Flying Corps began seeking flying schools in the U.S. for basic training. By the end of 1915, the Stinson School graduated five Canadians and Eddie Stinson. Around this time, Eddie mastered the controlled recovery from the aircraft tailspin that, up to that time, had killed many pilots. Eddie was then engaged by the head of the newly formed 1st aero squadron to teach the Army pilots training at the just-opened Kelly field near San Antonio.

Soon after, Eddie began a period of vagabond life, taking his undeniable flying skills wherever the opportunity to fly for pay arose. In the spring of 1917, he tested an airplane over Long Island, New York; in July, he was in Newport News, Virginia. Later in the year, he tested airplanes for the Curtiss Co. in upstate New York. The U.S. entered the war in April of 1917; Eddie Stinson reportedly applied for a commission in the U.S. Air Service that was denied when the required physical exam revealed probable TB in his lungs. Nonetheless, Eddie was taken on as a private and returned to Kelly Field, Texas, to serve as a flight instructor until the end of the war.

In 1919, just months after the war ended, Eddie bought five of the thousands of surplus Curtiss Jennies dumped on the market at nearly scrap prices. The planes were delivered to the port of Newport News where 25-year-old Eddie Stinson set up shop selling rides to the thousands of returning troops. Around that same time, former Lt. William A. Mara moved from St. Louis, Missouri, to Detroit, Michigan, to begin a new life. Mara correctly sensed that Detroit was in the early stages of boom-town explosive growth, and he wanted to be a part of it.

Stinson, meanwhile, had moved to Atlantic City to sell rides and while there, took part in various "aerial" promotional events, such as dropping eggs and mail by parachute. In the summer, he moved to Pittsburgh as the featured attraction in that city's effort to promote aviation. There he met fashion model Estelle Judy; they married in the fall of that year. The following year, the newlywed Stinsons joined younger brother Jack (shown above) in Dayton, Ohio, to assist him in organizing the "Stinson Aeroplane Co." and constructing an airplane out of existing bits and pieces, a forecast of things to come. After demonstrating the "Greyhound" just outside of New York City, on the way back to Dayton, the control stick weld broke, and with only rudder control, crashed—but Stinson survived with minimal damage to himself. The plane, however, was a total loss. Undaunted, he then moved to Birmingham, Alabama, where he again offered rides for a dollar.

This picture was taken the morning of December 29, 1921, on Roosevelt Field, Long Island, New York, just before the start of Eddie Stinson's attempt at a world record flight.

Stinson had moved to New York to test airplanes for the Aero Import Co., the U.S. distributor for the Italian Ansaldo Airplane Co. He delivered a six-passenger Ansaldo to El Paso, Texas, for eventual use in an air service between Mexico City and Juarez and, for a short time, was a Mexican general's personal pilot and provided rudimentary instruction to one of the general's staff. The affair ended badly when the "student" pilot crashed, and Eddie Stinson was ordered out of the country.

It was toward the end of 1920 that Eddie began to establish himself as "The Stinson" in aviation. Returning to New York, Eddie Stinson was hired by John Larsen, the importer of a crude looking but very good all-metal, German-built Junkers F-13, to test fly the airplane. With World War I just two years past, Larsen wanted to blur the craft's origin so he called it the Larsen JL-6. To help demonstrate the plane's ability, a JL-6 sporting auxiliary fuel tanks and Eddie Stinson at the controls took off from Roosevelt Field, Long Island, at 8:58:15 a.m. on December 29, 1921.

This picture of then world-record holder Eddie Stinson was taken Friday, December 30, 1921, after more than 23 hours in the air. Carrying co-pilot Lloyd Bertram and 360 gallons of gasoline, Stinson had set out to beat the world endurance record. After a grueling 26:19:35 hours in a freezing cold open cockpit, the plane landed, beating the world record by over two hours. The feat produced a three-page photo story in *Aviation Magazine*, the leader in its field, and similar comment in other reputable aviation publications. With that accomplishment, Eddie had clearly stepped ahead of his sister's position in the aviation limelight. (Photo courtesy Northville Public Library.)

While Eddie Stinson was setting records, W.A. (Bill) Mara was establishing himself in Detroit. With either luck or foresight, or both, he had secured a job at Detroit's prestigious Board of Commerce (BOC), then as much an upper echelon social club as it was a business-promoting organization. Mara had befriended Harvey Campbell, long-time vice president and secretary of the club, and had risen to assistant secretary of the organization. In that role, he had daily contact with the men who were at the core of Detroit's exploding business community. Mara was perfect for the job: a hard driver, usually in a three-piece suit with a handkerchief in the lapel pocket, his hair slicked back under a homburg hat, he was the straight-up man to see to get things done. So it was on the eve of 1922, that 27-year-old Eddie Stinson and 26-year-old Bill Mara were unknowingly about to have their interests and their lives mesh. Together, they would start a company that would build more airplanes for a longer period of time than any other Michigan airframe producer.

THE WINGED CITY OF DETROIT

In the Hotel Safe

This January 2, 1910 *Detroit News* editorial cartoon shows how Detroit had a long-standing interest in aviation. In 1922, the year Eddie Stinson arrived, Detroit had secured the national Pulitzer Trophy Air Races and acquired daily airline service to Cleveland. Additionally, Detroit was a boom town with money pouring in from the production of a million autos and a tidal wave of bootleg booze: strong attractions for anyone.

An ambitious young lady

An editorial cartoon in the October 16, 1921 issue of *Flying Magazine* depicts Detroit's dominance in the auto and speedboat industries and its desire for hegemony in airplane production. Eddie Stinson was sure to have seen it; perhaps it played a part in his decision to move to Detroit and tap into what appeared to be a market rich with aviation potential while his world record fame was still fresh.

One

The Principals Meet

Eddie Stinson arrived in Detroit sometime early in 1922. One of his first paying jobs was for W.B. Stout (who later gained fame as an airplane designer for Henry Ford) to test-fly his then-radical, all-metal Navy Torpedo Bomber. The airplane shown here was a daring gamble utilizing alloys and shaping techniques never before tested in the United States. With everything riding on the flight test, Stout chose Stinson to prove his airplane.

April 1922, the day of the big test, witnessed by Rear Admiral W.A. Moffett, Chief of Navy Aeronautics. Eddie Stinson is seen in the center and W.B. Stout (with the bow tie) to his right, next to the Admiral. Stinson had flown the plane successfully and suggested changes that were not made. A military pilot testing the plane crashed while landing. The plane was a total loss and a financial disaster for Stout. At that time, Stout was the head of the Detroit Board of Commerce Aeronautics Committee, Bill Mara was its Secretary, and Eddie Stinson was perceived as the best pilot in Detroit.

By the time of the Stout test, it is almost certain that Stinson and Mara had met. Clearly, Mara had an interest in aviation and part of his job was to promote it. Stinson was in town to make-a-go of the flying business. Their collaboration was a natural thing, and they liked each other even though they were vastly different. Eddie Stinson was a freewheeling, aerial phenomenon in coveralls and a tradesman's cap while buttoned-down Bill Mara was at home within the upper echelons of Detroit's business community.

The plane Stinson brought with him when he arrived in Detroit was a used Post Office German built Junkers F-13 (renamed the Larsen JL-6 for U.S. sales). It was the same type of airplane in which Stinson had set records just prior to going to Detroit. While crude looking, the craft was advanced in design and featured a thick, high lift, cantilever wing, and all-metal construction. Versions of the Junkers plane had been in commercial service in Europe since just after World War I. Imagineer/designer W.B. Stout used these same design concepts in his all-metal 2-AT airplanes, which impressed Henry Ford to such a degree that he bought Stout's company.

In May 1922, Eddie Stinson (fourth from the left) was hired for a Detroit-to-New-York flight by Charles Dickenson (on Stinson's left), the same man who sponsored the Chicago-to-New-York nonstop flight the previous fall. The flight, which must have been aided by a strong tailwind, was made in 2 hours and 32 minutes. Eddie Stinson's airplane looks a little worse for wear but made the flight in very good time. Bill Mara, who was then editor of the Board of Commerce *Detroiter* magazine, saw to it that a half-page picture of the event ran in the May 15 issue of the publication. It was the first of many Mara efforts to promote Eddie Stinson.

Eddie Stinson got another tough test-flying task in the fall of 1922. The plane to be tested was designed by M.E. Dare and featured a variable camber wing that he had patented in 1919. The idea was to have the pilot move a lever in the cockpit to mechanically increase the wing's camber for take-off and landing, then flatten the airfoil for better cruise performance. Eddie not only had to fly an untried airplane, he also had to learn—on the job—to manipulate the unproven camber controls. Eddie Stinson did both. The plane went through years of additional development and refinement; then, when it seemed near to perfection, a crash during a demonstration for the press killed Dare and ended the enterprise.

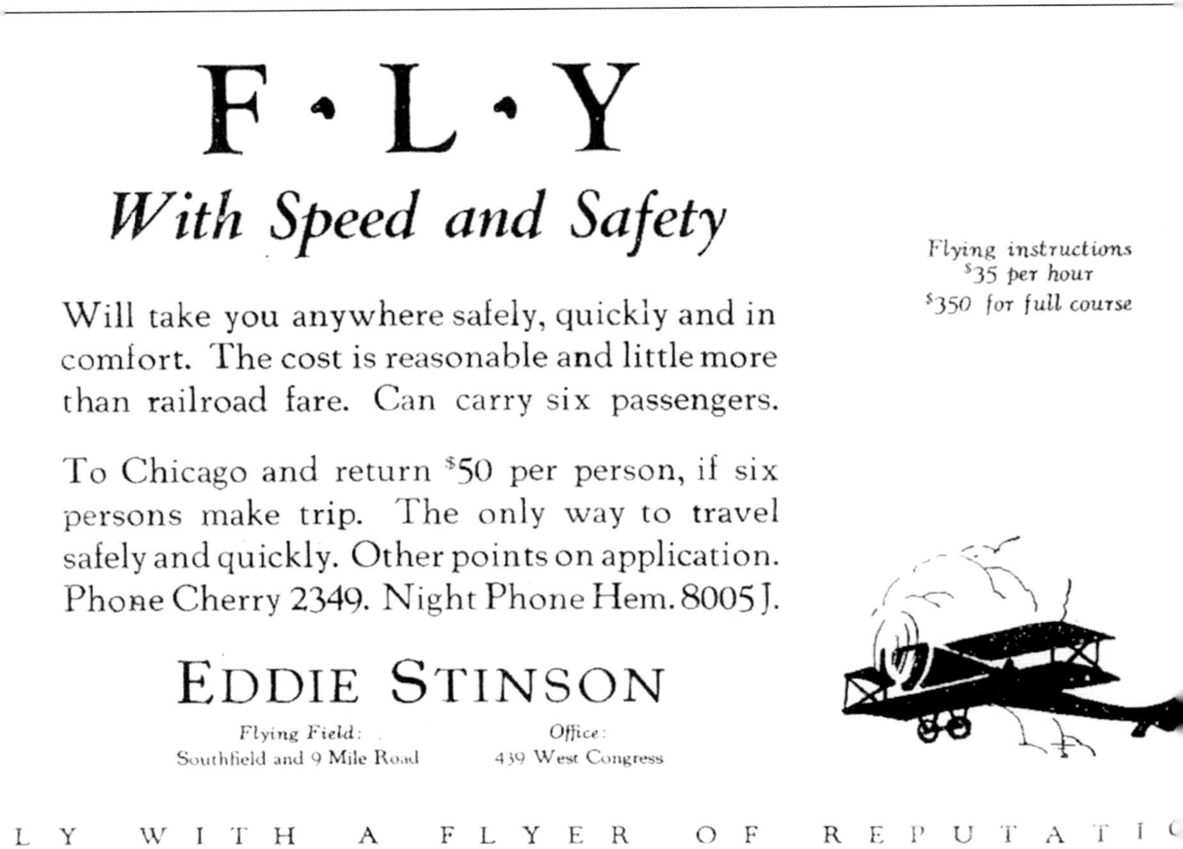

Firmly established in Detroit, Eddie Stinson sought to widen his economic base by offering flight instruction at $35 per hour or $350 for a complete course. Stinson, undoubtedly with Bill Mara's help, began running advertisements such as the one shown above in the city's premier business publication, the BOC's weekly *Detroiter* magazine. Note that the ad offers Detroit-to-Chicago round trips for $50 if all of the Junkers' six seats are filled. Also note the reference to railroad fares, an important consideration then in the pricing of air travel.

Eddie Stinson, with someone who appears to be a student pilot, in a World War I surplus "Jenny" training plane. The location is probably the suburban flying field mentioned in Stinson's ad, at Southfield and Nine Mile Road, now a highway interchange and an office complex in suburban Detroit. Detroit, in spite of a professed interest in aviation, did not have an airport at that time. The tail of Stinson's Junkers appears just under the nose of the "Jenny." At $35 per hour, flying lessons—then as now— were not cheap, but $350 for a complete course was a bargain, bearing in mind that there were no Federal Pilots Licenses issued at that time.

THE DETROITER

Interesting Detroiters — Eddie Stinson

To help his friend Eddie Stinson establish himself in the Detroit area, Bill Mara, who had been producing many stories about aviation in the *Detroiter* magazine, profiled Eddie Stinson in January of 1923 in a story titled "Interesting Detroiters." Eddie had a reputation of being uncommonly fond of bootleg liquid spirits. Bill Mara's story was a three-page piece that described Stinson in the most glowing terms, recounting his past achievements and adding his own testimony to the safety of flying with Eddie Stinson. That story, in a very respected magazine, written by its editor, was a priceless boost to the affairs of Stinson in Detroit.

Perhaps as a result of Mara's Stinson story in the *Detroiter* magazine, the Detroit BOC—aggressively promoting aviation in Detroit—became a frequent user of Stinson's services while conducting business around the state and beyond. Stinson is second from the left, Bill Mara is directly under the propeller hub, and Harvey Campbell, Mara's boss, is second from the right. It should be noted that at that time, Stinson's Junkers F-13 airplane was the only one in the area that could transport a group of that size. In spite of the plane's appearance, it was an efficient flyer and under Stinson's care, amassed a remarkable safety record.

THE DETROITER

Vol. XVI. Detroit, Michigan, September 28, 1925 No. 49

This is the new three-engine Fokker Air Liner, which is the largest ship in the Reliability Tour. It carries two pilots, eight passengers and 350 pounds of baggage at a speed of 120 miles per hour. The plane flies with a full load with two engines, the third being held in reserve as insurance against a forced landing.

Air Crusaders Off on History-Making Flight

Detroit was abuzz with aviation enthusiasm in 1925. Henry Ford had just entered the field, and Congress had just passed the Kelly Air Mail Act. Striving to promote aviation in Detroit, Harvey Campbell, the driving force of the BOC, gathered like-minded Bill Metzger, Bill Mara, and Lee Barrett. The foursome developed the idea to stage an airplane tour to cover 13 cities and 1,770 miles in 7 days. Bill Mara headed the project, Eddie piloted the pathfinder airplane, and Mara set arrangements along the route. The 17 entrants took off from Ford Airport on September 28 and generated national headlines.

On Sunday, October 4, 1925, the tour planes returned to Ford Airport in rain and fog. Nevertheless, thousands of people showed up to see the planes. Henry Ford, pleased with the promotion, delegated William B. Mayo, Ford's chief engineer and aviation booster, to have lunch with Mara to give him a check for $10,000 for his effort. Mara turned it down, saying the tour was simply his job. That decision proved beneficial down the road. Additionally, the Stinson/Mara pathfinder mission and the contacts made en route also had profound future implications for both men.

Two

The Beginning of Stinson Aircraft Operations

Bill Mara joined the prestigious Detroit Athletic Club (DAC) in 1925. His boss and friend, Harvey Campbell, was a long-time member and a sponsor. Mara already knew many of the members through his activities at the BOC. The DAC was, and is, the place where ideas take flight. Its membership contributed most of the funding for William Stout's Metal Airplane Company. Shortly after the 1925 Air Tour, Eddie Stinson showed Campbell and Mara an idea for a new type of airplane. They immediately thought of their friends in the DAC as potential investors for the prototype. Although not an engineer, Stinson, with his thousands of hours of flight and experience with wealthy passengers, had a keen sense of what the civilian market wanted. With Campbell's approval, Mara set out to raise $25,000 for a prototype airplane. He quickly gathered 25 men willing to invest (20 were DAC members, including the mayor of Detroit), created the Stinson Airplane Syndicate, and guaranteed to build the plane in three months' time. Eddie Stinson was named president and Bill Mara secretary of the syndicate.

Stinson's prototype airplane was ready for public demonstration by February 21, 1926. On that wintry Sunday, more than 70 people lined up for rides. The short demonstration flights took place at the Packard airfield on the east side of Detroit. Passengers rode in comfort in an enclosed, heated cabin at a time of year when most open cockpit airplanes were in winter storage. Named "Stinson Detroiter" by Bill Mara, the plane combined, for the first time, a heated, enclosed cabin, an electric starter, a modern air-cooled engine, wheel brakes, and luxury features such as an electric cigar lighter, ash trays, carpeting, and upholstered seats and side walls.

The wheel brakes were one of the most significant features of Stinson's new airplane. Until that time, airplanes were slowed on the ground by the tailskid dragging on the turf. With wheel brakes, a pilot could considerably shorten his landing roll and, thus, get into shorter fields. According to Bill Mara, Stinson used Harley-Davidson motorcycle brakes on the prototype. Note the tire chains: on the day of the public demonstration, Packard Field was snow-covered, negating the effect of the brakes. At the last moment, Eddie Stinson fitted automobile chains to the wheels to prove their effectiveness under the conditions that day.

Bill Mara, exercising his promotional skills, had sent sets of *Detroiter* pictures, copy, and specifications to the world's leading aviation magazines and was rewarded in the U.S. and England with multi-page spreads of the new-concept airplane. The specifications of the plane were as follows: Span 33 feet, 9 inches; Chord 6 feet; Wing area 350 square feet; Length 28 feet, Empty weight 1,700 pounds; Useful load 1,200 pounds; Fuel capacity 76 gallons; Oil capacity 7.5 gallons; Speed range 45-125 m.p.h.; Cruising range 500 miles; Engine 200 horsepower Wright Whirlwind. Stories about the Stinson Detroiter all focused on the plane's brakes and the far forward placement of the landing gear to avoid nose-overs when the brakes were applied.

The prototype airplane (built on time and under budget) was of conventional, welded, steel-tube construction with wood wing spars and ribs, and all fabric-covered except for the passenger cabin forward. According to notes written by Bill Mara, the prototype was sold to Horace Dodge, son of one of the Dodge Motor Car founders. Some records indicate that H. Dodge bought a later model but Mara's notes are clear; perhaps Dodge bought two Stinsons. The prototype Detroiter plane design was altered drastically when the Stinson Syndicate, having met with some success, decided to incorporate and go into the airplane production business.

The Stinson Detroiter generated great interest in the aviation industry. Among the early viewers of the airplane was Capt. Eddie Rickenbacker, America's ace of aces in World War I. A few years earlier, Rickenbacker had lent his name to an auto company, and Eddie Stinson was hired to fly him around the country to sign up dealers. The trip ended badly when Stinson's Junkers crashed. Rickenbacker was to later head Eastern Air Lines. There seemed to be no hard feelings when this photo was taken.

Visitors came from afar to see the Detroiter. Here with Eddie is Major General Sir Sefton Brancker, Director of Commercial Aviation for the British Empire. General Brancker was quoted as saying, "The Stinson Detroiter airplane (was) the most interesting development (I have) seen in that it is absolutely stable and almost flies itself." England would later be an important market for Stinson and received many Lend Lease airplanes during World War II.

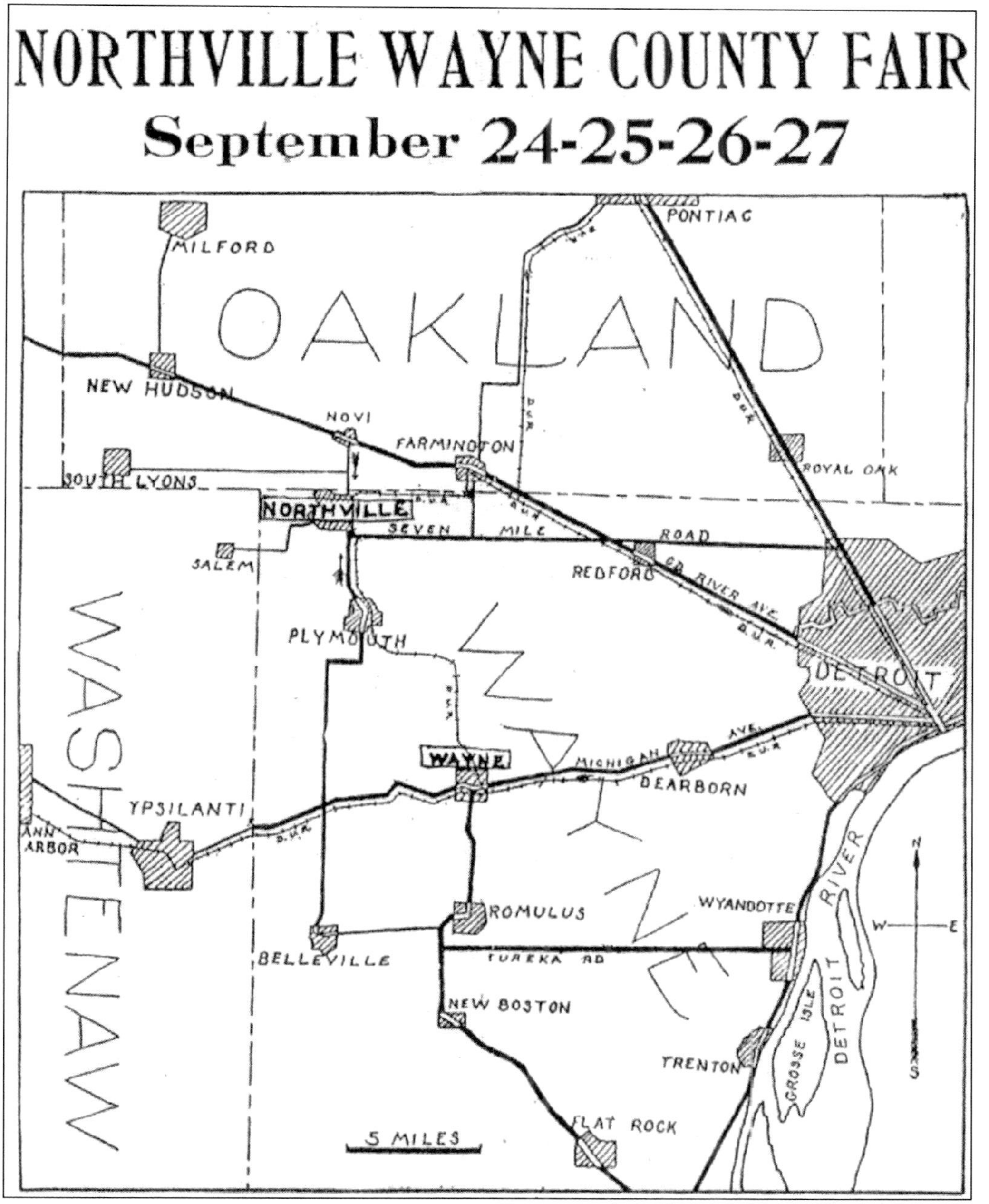

In May of 1926, the Stinson Syndicate became the Stinson Aircraft Corporation with most of the original investors buying $150,000 worth of shares. Production was to be in Northville, Michigan, a town of 2,500 located about 30 miles northwest of the downtown Detroit loft where the prototype was built. Though distant, the Northville site had the advantages of a railroad, an interurban line to bring in workers, and an old, empty factory owned by one of the investors. There was also open land for a landing field, held by cooperative Northvillians. Note that on the map the three places Stinson had operations, Northville, Wayne, and Belleville, are all located in close proximity.

Stinson production of the revised prototype then identified as the SB-1 began in August of 1926. The factory was technically a New England industrial style, two-story building with a roof monitor built of brick in the 1890s. It measured 50 feet by 194 feet, and had over the years been a furniture and a scale factory. Because of the interior floor supports, it was impossible to completely assemble airplanes inside the building. Final assembly took place after the planes, minus wings, were towed through town to the landing strip a couple of miles to the west.

Records indicate that 26 Stinson SB-1s (meaning Stinson Bi-Plane 1) were built in the Northville factory. The first registered plane went to the Aeronautics Bureau of the Department of Commerce, Washington, D.C. The second plane, shown here, clearly illustrates the simpler shape of the production model versus the prototype. This plane was sold to the Dodgeson Motor Co. Differences from the prototype were: a wing span increase of two feet, a chord increase by 2 inches, and the length by 10 inches. Even though SB-1 was being sold and registered in the fall of 1926, it did not get its Federal Approved Type Certificate until January of 1928, when it was essentially out of production.

The Dodgeson Motor Co. logo on the fuselage was the emblem of John Duvall Dodge's attempt to enter the auto manufacturing business with a car of his own. The son of John Dodge and cousin of Horace Dodge, he was well-positioned to raise money for a new enterprise. While the SB-1 was useful in traveling the country to raise funds, there may well have been a bit of one-upmanship in the purchase of a later-model airplane than his cousin Horace had. It is also remarkable that Stinson airplanes had so quickly been adopted by one of America's richest families.

This heavily retouched photograph is one of the very few showing the instrument panel of the Stinson SB-1. The retouching highlights the details of the rudder foot controls and their absence for the co-pilot, the stabilizer trim lever at the left, looking very much like the emergency brake handle of the cars of the day, and the automobile-type steering/control wheels. Engine and flight instruments, while few in number, are large in size. The featured electric cigar lighter appears on the left, with ash tray on the right.

Another retouched photo shows what, for that time, was considered a luxurious airplane interior. This picture appeared in an SB-1 ad but is most likely a photo of the prototype airplane, as indicated by lines in the rear roof. The prototype had a unique fuselage shape in the rear of the cabin area not carried on in the SB-1. With entry doors behind the pilot's seat, it must have been a tight squeeze between the seats for the pilot to get into position.

During the Air Tour pathfinder flight the previous year, Bill Mara had met Col. L.H. Brittin in St. Paul, Minnesota. In the fall of 1926, Brittin had an opportunity to get an Air Mail route between St. Paul and Chicago, but he had to act fast. After trying unsuccessfully to get other airlines interested, he decided to start his own. Mara, trying to sell airplanes, developed a $100,000 financial plan for him that included three SB-1s at $12,500 each. Brittin went to Detroit with $50,000 in pledges. Mara gathered many of the DAC members who were Stinson investors, and they matched Brittin's funds. As a result, Northwest Airways was incorporated in Detroit September 1, 1926. Three of the first four operating officers were DAC members. Seated are F.W. Blair, Vice President, and H.H. Emmons, President. Standing are W.B. Stout, Secretary, and Col. L.H. Brittin, General Manager.

This photo, taken inside the Stinson factory in the fall of 1926, shows one of the three Northwest Airways planes nearing completion. The fuselage fabric is on and the engine hung. Pictured from left to right are Capt. Ray Collins, who worked for Stinson's Vice President F.W. Blair; Eddie Stinson; Bill Mara; and Northwest Airways founder L.H. Brittin. Except for Eddie Stinson, the group looks pretty glum. However, the SB-1s performed sc well that Northwest went back to buy many Stinson airplanes.

One of the three Northwest Airways SB-1s is pictured, fresh from the factory, probably on the Stinson/Northville landing field. The plane has the Northwest Airways lcgo, barely discernable on the fuselage, but was at that time without its registration numbers. Note the huge landing lights under each lower wing. They were installed to help pilots find, and land at, the then-unlighted airports.

November 1, 1926 was delivery day for the Northwest Airways fleet of Stinsons. Preparing to depart, in very poor weather, are, from left to right, Pilot Dave Behnke (who later would head the pilots' union), "Shorty" Schroeder, who as Major serving at McCook Field set a world altitude record of 33,000 feet in 1920 and later managed Henry Ford's airport operations, and Col. Lewis H. Brittin, founder and general manager of the airline. It should be noted that, at that time, there were virtually no paying airline passengers. The lines survived on the income generated by carrying air mail. Northwest Airways sold its first passenger ticket in July of 1927, nine months after receiving its Stinson SB-1 airplanes.

Adding to the impact of the SB-1's editorial coverage, Mara launched new ads in May of 1927. In spite of being new in the market, Mara was able to point to three airline purchasers, one car company, the Department of Commerce, a famous explorer, several air taxi services, two auto scions, a major newspaper, a famous airplane designer, and other noteworthy individuals. All told, Stinson had made 18 SB-1 sales, some multiple units, in seven months. To be sure, the Stinson name carried great weight, but Bill Mara should be given credit for adroitly handling the public relations and advertising aspect of the launch.

Toward the end of 1926, Eddie Stinson began development of a monoplane. Times were changing: President Coolidge had signed the Air Commerce Act federally regulating civil aeronautics, in Wichita, Kansas, Clyde Cessna was experimenting with his own monoplane, and Richard Byrd had flown over the North Pole in a Fokker monoplane. In his role as sales manager, selling what he had, Bill Mara began a series of SB-1 advertisements in national aviation magazines in 1927. The advertisements asked potential customers to compare and test ride, essentially asking them to buy an airplane as they would buy cars.

While the SB-1 was not built in large numbers, many of its purchasers performed feats that kept the airplane in the headlines. Capt. George H. Wilkens is pictured second from left, talking with Eddie Stinson, with pilot Ben Eielson and Bill Mara at right. Supported by the *Detroit News*, Wilkens bought an SB-1 registered as NC 5262 as a support plane for his attempt to discover land at the North Pole. This photo was taken inside the Henry Ford hangar in Dearborn, Michigan. It is very likely that Ford allowed use of his facilities as payment for Mara's work (for which he refused to be paid) on the previous year's Ford Air Tour.

Wilkens' Stinson performed well during the expedition. The purpose of the mission was to explore the North Pole looking for land with the idea of claiming the land for the U.S. The Detroit area had a large investment in the Wilkens adventure. There was the support of the Aviation Committee of the Board of Commerce, the Detroit Aviation Society, the sponsorship of the *Detroit News*, there were the general good feelings toward Eddie Stinson, and finally, there was a city-wide campaign led by the mayor to raise funds. As part of that campaign, school kids contributed their change to fund the event. No land was found.

The SB-1 proved popular with Canadian air operators where the planes' ruggedness was imperative. Of the 26 built, four were put in Canadian service. Bill Mara and Eddie Stinson, at the right, are shown here delivering the Canadian Air Express company's airplane Number 1. The Toronto-based plane carried Canadian registration G-CAFW. It was the next to last SB-1 to be built. Even as this airplane was being delivered, work on Eddie Stinson's monoplane, to be designated the SM-1, was moving forward.

Florida Airways, like Northwest Airways, was among the first to get Air Mail routes and to use Eddie Stinson's SB-1 airplane to service its routes. Florida Airways was co-founded by World War I air aces Reed Chambers and Eddie Rickenbacker. Rickenbacker knew Eddie Stinson well and had examined the prototype airplane. The award of CAM 10, which was the Air Mail route from Atlanta to Miami, started their operations on April 1, 1926. In spite of the high profile of the founders, the airline failed a few years later.

Always on the lookout for a promotional opportunity, in July of 1926 Bill Mara arranged to give the Governor of the State of Michigan his first airplane ride. In the center of the photograph are Mara and Stinson; at Eddie Stinson's left is Governor Groesbeck, and at either side of them are officials from the State Chamber of Commerce and Detroit and Lansing Banks. It was a great time to be in Detroit. Car and real estate sales were booming, Air Mail service had arrived, and the city fathers were talking seriously, finally, about establishing a municipal airport.

Edward F. Schlee bought several Stinson SB-1s for his Wayco air taxi service. The first sale came about when Schlee called to have Stinson fly him to Mackinac Island, a summer resort island between upper and lower Michigan, that to this day does not permit gasoline-propelled vehicles. Mara agreed to the flight if the wealthy Shell Oil distributor would agree to buy an airplane, which he did. On July 28, 1926, Eddie Stinson landed on the golf course in front of the Grand Hotel, taxied near to the lobby, and deposited Mr. and Mrs. Schlee in time for dinner. The event was a huge promotional success for Stinson.

In the spring of 1927, Stinson's instinct about airplanes was proven again. During May 20 and 21 of 1927, Charles Lindbergh flew nonstop from New York to Paris in a Ryan monoplane (shown here in Ford's hangar) and exploded the aviation world. At that time, Stinson's monoplane the SM-1 was in final tests. Although developed independently, the Ryan and the SM-1 were quite similar. Compared to the Ryan "Spirit of St. Louis," the SM-1's wings were 2 inches shorter, with identical chord, and it was 4 feet 8 inches longer. Both used a Wright 220 horsepower engine and were within a couple miles of each other in cruising speed. The Spirit of St. Louis cost $10,580, the SM-1 $12,500; the SM-1 had wheel brakes, the Ryan did not.

The first SM-1 production unit produced was sold to the aforementioned Edward Schlee and proved be to an instant winner. On June 27, Eddie Stinson took off on the third Ford Air Tour, piloting Schlee's new SM-1, named "Miss Wayco." Sixteen days and 4,000 miles later, Stinson and the SM-1 won the Tour by 2,000 points over the nearest competitor. The plane, packed with Mrs. and Mr. Schlee, daughter Rosemarie, Messrs. Klemmer and Plank, Bill Mara and Eddie Stinson, carried the heaviest load of 14 competitors. Paul Braniff, with whom Mara would later engage in business deals, finished last.

Winning the Ford Air Tour generated instant fame for the SM-1. It was a time of aviation record-making. In June, Chamberlin and Levine flew nonstop New York to Berlin, then Lts. Maitland and Hegenberger flew nonstop Oakland, California, to Hawaii. Ed Schlee, shown above with Billy Brock, decided to take his SM-1, renamed the "Pride of Detroit," on a globe-circling flight. Taking off in August, the pair flew east, successfully covering two-thirds of the world before deciding to end the flight in Tokyo. Their flight heaped more recognition on Stinson's airplane. That plane is on display at Henry Ford Museum in Dearborn, Michigan.

Attempting to feed on the success of the SM-1, and perhaps to curry favor with the purchasing influences in the Stinson organization, the Consolidated Instrument Company ran testimonial ads in the aviation press. Their message to the aviation audience was to link the usage of their product to the recognized prestige of the Stinson brand, thereby enhancing their own position in the marketplace. At that time, the products of the Consolidated Co. were not market leaders.

Another supplier trying to reach the aviation market was the Berry Brothers Paint Company. They were important because Bill Mara agreed to paint a Stinson airplane any color the buyer wanted. In addition to a glowing testimonial from Eddie Stinson, the ad informs that wing wood and metal parts get two coats of "Lionoil" (paint), the fuselage steel gets red primer and Berryloid (paint), the fabric covering gets five coats of Nitrate Dope plus two coats of paint (any color desired), and the Duralumin cowl gets one coat of primer and two coats of paint. This ad serves as a strong statement about quality finishing applied to Stinson airplanes.

During the warm months of 1927, the Stinson organization had a hangar erected at their landing field. Planes still had to be dragged, without wings, through town, but at least the final assembly could take place in reasonable comfort. That hangar also served as a service site, a function of growing importance as the fleet of Stinson airplanes was growing rapidly. SM-1 sales had been good and orders strong; the time was ripe to focus on production and delivery.

All sorts of suppliers to the Stinson organization wanted to benefit from associating with a winner. While there were no huge cash benefits for the subject of the ad, there often were price discounts on the subject product. The cement floor of this hangar was still visible a few years ago—the landing field/hangar area is now an up-market housing development. Newspaper reports of the field's construction mention the use of the nearby Detroit House of Corrections labor force to remove stones.

Three

Building the Stinson SM-1 in Northville

The Stinson Corporation's first year had gone extremely well. The SB-1 product had been good, and they were ideally positioned to take market advantage of the Lindbergh boom with their SM-1 monoplane. Bill Mara's advertising and product exposure efforts had been on target and his efforts to gain distributors had been going well. The record-setting exploits of their customers also built interest in the product, and Eddie Stinson's and Bill Mara's personal selling of airplanes had generated orders. They now had to tackle the largely unknown problems of building a modest volume of airplanes in a less-than-satisfactory factory.

The SM-1 cockpit control area was very similar to the SB-1 and very simple by today's standards. But in 1927 there were no regularly operating aircraft radios, navigation aids, or blind flying instruments. The round portholes just under the windshield were intended to increase downward visibility when landing on the grass fields of the day. Just under the control wheels are the heater outlets for the pilots. The bright spots in the pilot's rudder pedals are wheel brake controls. Apparently, pressure in the left pedal pads could apply simultaneous braking, while pressure on just the bottom pads could produce independent braking. The co-pilot had no brake control. According to NACA Aircraft Circular No. 60, the plane had non-shatter glass throughout.

Six wicker chairs, with leatherette slipcovers, provided seating in the SM-1. Wicker chairs were very common for multi-passenger airplanes of the day, from the mighty Ford Tri-Motor on down. They were light, cheap, reasonably strong, and comfortable, but could not pass any test today. The back seat, shown above, is typical of wicker seating of the day. The SM-1 had a six-*inch* aisle between the seats for movement within the cabin. In the rear cabin wall there was a door that gave access to a storage compartment. There were doors, not visible in this photograph, on each side of the cabin; they were wood-framed, as were the windowsills. With carpets on the floor and upholstery on the side walls and ceiling, it is clear at this early stage that Stinson was building airplanes with appeals to the quality/luxury market.

The wings nearing 46 feet in width were built of spruce and were fitted with welded steel tubing ailerons; there were no flaps. Each of the nearly four dozen wing ribs was made up of approximately three dozen pieces of square spruce strips. Each joint was carefully fitted, glued, and reinforced with glued and nailed plywood gussets. The finished ribs were then slid into position along the front and rear spars. The leading edge of the wing was covered with aluminum, then the entire wing was covered in fabric. The factory roof monitor, seen at the top of the photo, indicates the wings were built on the second floor. Note the simple sawhorse jigs and the lack of supplemental lighting. Also, see the set of pontoons at the left of the picture; they added $2,500 to the $12,500 price of the Stinson.

This is another scene taking place on the 50-foot by 190-foot upper floor of the Stinson factory. Here a wing is having "pinking" tape applied over the thread stitching that held the fabric to the wing ribs; the fabric was not glued to the wood. The pinking tape was adhered to the wing with "dope"; note the bucket of dope on the wing. In the background, workers are applying dope with brushes to tighten the fabric. The wing would get several applications of dope to shrink, tighten, and waterproof the fabric. The drying time of the doped wings varied with temperature and humidity, there being no drying ovens in the Stinson factory. Each wing contained a fuel tank that drained into a 1.5-gallon auxiliary tank behind the firewall. Their purpose was to provide an ample supply of engine fuel no matter the attitude of the airplane.

Here the bottom side of a wing receives a finish coat of paint. It is a scene that would drive OSHA crazy today. The only illumination was from a couple of incandescent light bulbs burning in a spray paint atmosphere. And, the workers are not wearing spray masks, although they do have Stinson coveralls. Things were different in 1927, and "Stinson" was considered a good place to work. In fact, many of the 200 Stinson workers traveled many miles on the interurban trains to reach the Northville factory. It should also be remembered that neither Eddie Stinson nor Bill Mara had any experience in the management of volume airplane production. They, like most of their workers, had to learn the job while doing it. Note the two protrusions near the right end of the wing. They are fittings for the gasoline lines that carried the gravity-fed fuel to the engine. Note, also, the extra-heavy pinking tapes that outline the position of the gas tank inside the wing.

Pictured here is a sub-assembly welding location on the factory main floor fabricating what seems to be wing lift struts. Note how the workbenches are located along the outer wall as close as possible to the windows. There seems to be almost no supplemental lighting, although what appear to be electric wires are to be seen. One wonders how production fared during the winter months with its considerably fewer daylight hours. Also, there is no evidence of any portable fire extinguishers in this photo. The worker in the foreground has appropriate welding goggles, but he, like the others seen, is not wearing gloves. In Bill Mara's personal remembrance notes, he mentioned that Eddie Stinson, having once crashed as a result of an in-flight failure of one of his welds, was particularly insistent that all Stinson airplane welders and the resulting welds be of top quality.

The main floor of the factory, with a fuselage on the floor, probably just out of the welding jig shown in the bottom left. In the center of the photo, a worker appears to be adjusting the flame on his welding torch, probably to touch up some tubing joints. Note how the vertical pillars supporting the second floor make completely assembling an airplane inside the factory virtually impossible. Additionally, since the building was only 50 feet wide and the wingspan of an SM-1 was nearly 46 ft., it would have been a very tight fit. As can be seen, the only light seems to be coming from the side windows. It is quite certain that the building was wired; in the foreground there appears to be a bench-mounted, electric motor-powered grinder. Why the lighting was not better is unknown. It took dedicated workmen to build a quality product under these conditions.

Pictured here is an SM-1 fuselage nearing completion with two more close behind. The plane, still without its landing gear, is sitting on some sort of dolly. The fabric is on and the Wright engine is mounted. In the Northville Historical Society's oral histories, there are several mentions of the wooden engine crates being used as outdoor toy houses by generations of Northville children. An inspector seems to be examining a part in the lower left of the picture, holding it up to the window light to get a better view. Even though the inspector is wearing Stinson coveralls, he is also wearing a white shirt and tie. The picture on page 35 shows President Eddie Stinson in the same sort of dirty coveralls. Authentic Stinson coveralls have become a sought-after collector's item in the Northville area.

This is a detail photo of the SM-1 welded steel empennage. This was probably taken behind the factory next to a storage shed, providing more evidence of the cramped conditions at the factory. The photo also reveals the close spacing of the empennage ribs and the sturdy rudder post. Barely apparent is the tail skid. It was about this time that the tail wheel was patented by the Ford Airplane Division and made free for others to use. Stinson was soon installing tail wheels.

This is another detail photo showing the cockpit area. Clearly shown is the control wheel mechanism linkage and the sturdy size of the components. The rudder pedals have not yet been installed. What appears to be an automobile emergency brake handle is the horizontal stabilizer trim control lever with the push-pull rod below and running to the rear. Stinson, like many of the airplane builders of that time, freely used automobile components when he could.

Here is a close-up of the SM-1 engine mount. With the cowling removed, it is easy to see the filler cap of the 7-gallon oil tank. Just behind the filler tube is the flexible metal tube that supplied heat to the cabin area. According to NACA circular No. 60, the engine is secured by a tubular steel mount mated to the fuselage by four bolts. To lessen vibration, "oil-soaked plywood washers" were installed between the engine and the mount.

The landing gear on early SM-1 models was essentially a carry-over from the SB-1. As is shown above, it was a split axle type with a bungee cord central shock absorber. Later SM-1 models featured an "outrigger" landing gear with hydraulic shock absorbers. The tubular landing gear struts were heat-treated to a strength of 160,000 pounds per square inch. Note the heavy fixed entrance steps; they would be a Stinson design feature for years to come.

The year 1927 launched a period of rapid advances in aviation. In Kansas, Clyde Cessna was getting into stride with his cantilever (no outside struts) winged airplane of conventional construction. On the west coast, Lockheed's brilliant Vega, which heralded acceptance of cantilever wings and monocoque fuselages, was just appearing. Over that winter, Eddie Stinson, anticipating an evolving market, developed a low wing prototype, its only picture shown above. It was a strut-braced, wide-tread airplane that probably used many SM-1 components. It is also likely the plane was built up in the landing field hangar, where this picture was taken, since space in the factory was so limited. Unfortunately, the craft did not fly well, and the idea was abandoned.

Early in the production of the SM-1, Bill Mara sold Paul Braniff, a former Ford Air Tour competitor, the above airplane without a deposit. When Mara delivered the plane in Oklahoma City, he found Paul Braniff had no money. Mara then went to his brother Tom Braniff and sold him an idea of a flying club, then gathered Oklahoma City backers for the idea, which later evolved into Braniff Airlines. It was the sort of can-do chutzpah Eddie Stinson and Bill Mara possessed that then characterized the Stinson organization.

Four

UPWARD AND ONWARD FOR STINSON

The SM-1 was an ideal airplane for the aviators of 1927 who were trying for rub-off glory from public interest in long-distance flights heightened by Lindbergh's feat. The plane's big wing could carry a heavy load, as could the landing gear; the Wright engine was well proven and the long fuselage could be fitted with additional, long-range fuel tanks. Not every adventurer was as fortunate as Brock and Schlee, who bought SM-1 No. 1. Pilot Paul Redfern bought No. 3 and attempted to fly nonstop from Brunswick, Georgia, to Rio de Janeiro and was never seen again. Another early SM-1, perhaps No. 6, named Sir John Carling, was piloted by two Canadians who attempted to fly nonstop from London, England, to London, Ontario, Canada (not far from Detroit), but disappeared without a trace. An attempt of a Windsor, Ontario, (across the river from Detroit), flight to Windsor, England, was abandoned as too risky. Pilot George Halderman and attractive Miss Ruth Elder in another early SM-1, named "American Girl," attempted to make her the first woman to fly the Atlantic. Taking off from Roosevelt Field, New York, they flew for 24 hours before a broken oil line forced them down, luckily landing next to a ship, where they were rescued and generated vast publicity. George Halderman soon after became a friend and sales agent for Stinson. In 1927, at least three publicity-seeking SM-1 flights ended without success but did not tarnish the reputation of Eddie Stinson or his SM-1.

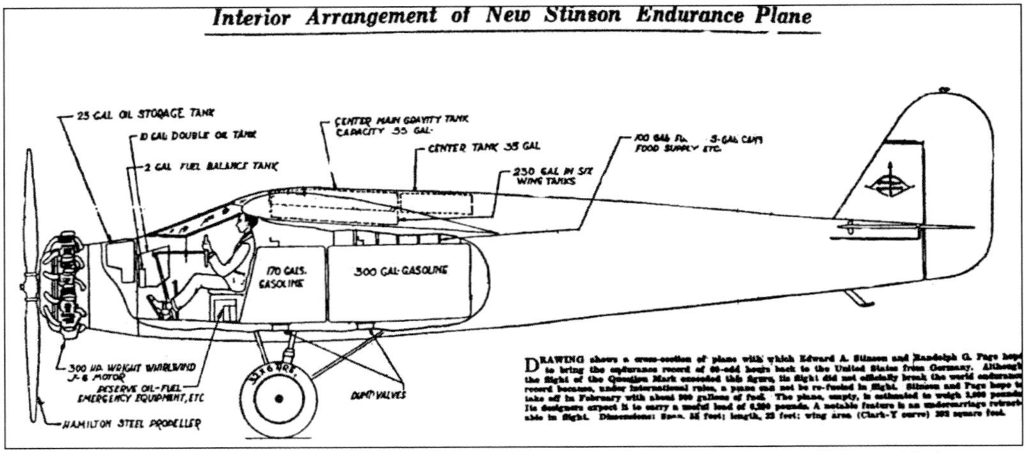

Perhaps driven by the rash of failed record attempts, Eddie Stinson set out to establish a new world endurance record early in 1928. This was done in an SM-1, registered as X1524, modified as generally depicted by the *Detroit News* on January 20, 1928. Early in March, Stinson began an attempt over Lake St. Clair but was thwarted by the "choppiness" of the air. On Wednesday, March 28, at 7:37 a.m., Eddie, with new friend George Halderman, took off from Jacksonville, Florida. For the next 53 hours and 37 minutes they flew a 30-mile course over Jacksonville, coming down on Friday night at 11:40 p.m. They beat the old record by more than an hour and collected $5,000 from the Chamber of Commerce for doing so.

Eddie Stinson, Estelle Stinson, and George Halderman are pictured in Jacksonville, Florida, the site of their victory. A few days later, the people of Northville packed the Presbyterian Church hall for a banquet to honor the flyers, who had momentarily captured the attention of the world. Among the speakers was Mrs. Stinson, who told of riding in the message plane that was used to deliver weather reports on printed signs to the record breakers. Such was the state of communications before radio.

The Stinson/Halderman successful endurance flight was badly needed. Highly publicized attempted long-distance flights that ended in disaster, like Paul Redfern's depicted here, could have brought strong negative feelings toward Stinson airplanes. The Stinson organization was out to sell airplanes; it had no control over what buyers did with them. However, the buyers' sometimes-foolhardy failed attempts to gain fame could have caused doubts about Stinson airplanes with disastrous results on sales.

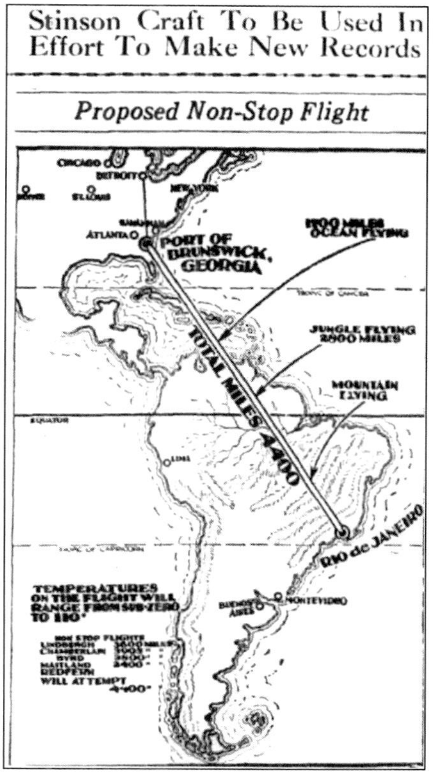

Here is Eddie Stinson's record-breaking SM-1, back at its home base at the bleak and snowy Northville airfield. Not discernable in this picture are the signs painted on the fuselage announcing that this plane set a world endurance record. That flight generated a massive amount of positive publicity in newspapers and in the leading aviation magazines: *Aero Digest*, *Aviation*, and *Air Transportation*. That publicity, in turn, solidified the perception that Stinson airplanes were good and that failed endurance attempts were probably not the fault of the SM-1.

While sales of the six-passenger SM-1 were good, it was apparent that Buhl, Cessna, and Ryan (still touting its "Spirit of St. Louis" success), as well as others, were having their own success with slightly lower-priced passenger cabin airplanes of the type Eddie Stinson had pioneered. To meet the competition while maintaining Stinson-quality standards, a smaller, four-passenger plane was designed, utilizing a smaller, 120 horsepower locally-produced Warner radial engine with a lower price tag.

The new model, "Stinson-Detroiter Junior," SM-2 was first publicly shown at the Detroit All-American Aircraft Show in April of 1928. According to published reports, Stinson Chief Engineer William C. Naylor said the plane was designed in three weeks and built in three weeks. It was a feat impossible to duplicate today.
In addition to being smaller, the Stinson Junior departed from tradition by replacing the tail skid with a tail wheel and adopting the stick-type control instead of the traditional Stinson wheel.

This 1928 Stinson SM-2AA, photographed at the Antique Airplane Association Airfield at Blakesburg, Iowa, is probably the oldest Stinson flying today. When the SM-2 first appeared, it was a sensation. It looked like its bigger, older brother, but it was smaller in every dimension. The wingspan and length were about 6 feet shorter, it was about 1,000 pounds lighter, the horsepower was cut in half, and, importantly, the price was nearly cut in half. The published price for the Warner engine model was $6,750 compared to $12,500 for the SM-1 with the Wright J-5 engine.

While the SM-2 had a lower price tag, it had what had come to be known as Stinson quality. The windows were shatterproof glass, the front two could be rolled down, there was a new skylight over the control area, luggage storage under the seats was provided, and the interior was trimmed in leather matching the seat upholstery. In performance, the SM-2's big brother could carry more at a faster speed but was nearly twice the price; the SM-2's, top and cruise speeds were about 80 percent of the SM-1, and loads were about 75% percent. The SM-2 was still a big airplane, compared to the SM-1, which was considered huge.

From this view it is difficult to tell an SM-2 from an SM-1. In terms of dimensions the SM-2 had a wingspan of 41 feet 6 inches, the overall height was 7 feet 9 inches, and the length was 26 feet. Wing-mounted gasoline tanks carried a total of 45 gallons. The Warner 120 horsepower engine burned 6.5 gallons per hour at a cruising speed of 95 miles per hour. Range was estimated at 600 miles with an endurance of 6.9 hours. Empty weight was 1,428 pounds and gross weight was 2,253 pounds.

This view of NC 8471 at the Antique Airplane Association Airfield provides a glimpse of the environment in which this airplane was born. The diagonal paint treatment on top and bottom of the wings became a standard Stinson design. Fuselage striping varied over the years as did the tail fin Stinson logo design. One sure way to tell an SM-2 from an SM-1 is the lack of the auxiliary fuselage side windows under and ahead of the windshield.

The SM-2 instrument panel was a no-nonsense Spartan information center. It was simple and small in area, because in 1928 instruments were few and radios and navigation aids were non-existent. The uncrowded instruments were large and easy to read and were lighted from the front rather than the back. The throttle knob, in the center of the panel, looks the size of a tennis ball and is easily reachable from either seat. In the foreground are the stick controls, a design departure from previous and future Stinsons.

This landing gear detail photo also shows the style of 1928 landing lights. It was designed to compensate for the rough terrain a Stinson Junior was bound to encounter on the landing fields of the day. The entire landing chassis, including the tail wheel, was stressed for a load of 12,500 pounds. The tail wheel was fitted with a 12 inch by 3 inch tire, large by today's standards. The standard main wheels were 28 inches by 4 inches. An optional tire of 30 inches by 5 inches was available for soft fields.

Stinson had sold a number of airplanes to the Department of Commerce. This picture shows a high-level group from the Department visiting a high-level group from the Stinson Corporation. Pictured from left to right are Bill Mara, Secretary, Eddie Stinson, Col. Clarence M. Young, Assistant Secretary of Commerce, Aviation, Capt. John F. Donaldson, and A.P. Tappan who had served under Col. Young in World War I. The year 1928 was a good one for sales, and production was pushing the capacity of the old Northville factory to the limit. That and other events just over the horizon were building up that would impact the lives of Eddie Stinson, Bill Mara, and the Stinson Airplane Corporation.

Not only were sales booming domestically, they were also doing well internationally as a result of favorable publicity and widespread advertising. Several SM-1s had gone to Peru; then the Mexican Government ordered six. Pictured above is Eddie Stinson in conversation with Col. Manuel Rojos from the Mexican military forces. One can only speculate on how good Stinson's Spanish was or how good the Colonel's English. The Colonel is in Class A uniform for the occasion, including dress gloves.

Another Mexican delivery, the photo reads, "Ready for Mexico, Stinson in front of the Stinson Junior just before the start. Sept. 14, 1928." The location is the Stinson flying field just outside of Northville. About 40 SM-2 Juniors were built in 1928. Even though international sales were often government sales, the Stinson organization seems to have made virtually no effort to approach and sell to the military side of the U.S. Government.

AIR TRANSPORTATION

THE TRADE JOURNAL OF COMMERCIAL AVIATION

...ed Weekly at 34 N. Crystal St., East Stroudsburg, Pa. Executive and Editorial Offices at 1265 Broadway, New York C...

... No. 8 SEPTEMBER 8, 1928 15 CENT...

BLAZING THE TRAILS IN MEXICO . . . .

Six gigantic birds — whirring over tropical wastes and snow-capped peaks — start soon to blaze the air mail trails in Mexico.

Stinson-Detroiter Monoplanes were chosen by the Mexican government for this pioneering because the most rigid tests proved beyond a doubt that *they carried greater pay-loads at a higher degree of efficiency than any other aircraft known.*

The Mexican government is but one of the many organizations to arrive at this conclusion. Stinson planes, known the world over for their ruggedness, reliability, economy and dependability, finished well ahead of all other single motored cabin planes in the Reliability Air Tours of 1927 and 1928.

Official performance records prove Stinson's leadership!

PRICES

Stinson-Detroiter, six seater . . . $12,500

Stinson Junior, four seater 6,750

At The Factory

STINSON AIRCRAFT

"The Aircraft Standard of the World"

Stinson Aircraft Corp. Detroit, Mich. U.S.A.

Stinson attempted to capitalize on the sale of airplanes to Mexico by placing a full-page front cover ad on *Air Transportation* magazine timed to coincide with the flight to Mexico. *Air Transportation* was a magazine well read by the people in the aviation business, the kind of audience that could influence the purchase of multiple units. The first line of copy stresses that six Stinsons would be carrying Mexico's Air Mail.

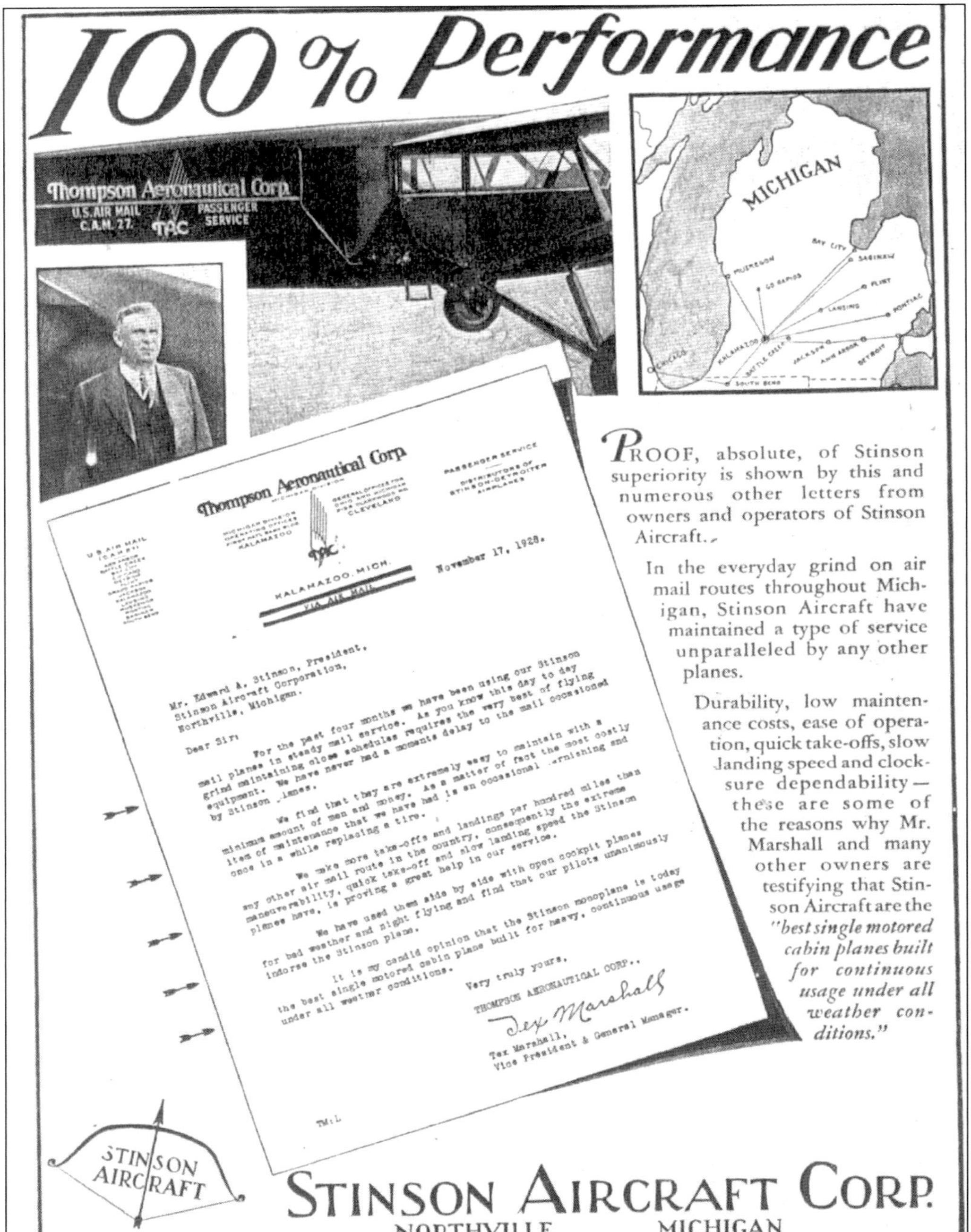

In December of 1928, keeping up the advertising pressure, Stinson ran a testimonial ad from the Thompson Aeronautical Corp., an Air Mail Carrier based in Kalamazoo, Michigan. In a very formal style, Tex Marshall, Vice President and General Manager, said, "It is my candid opinion that the Stinson monoplane is today the best single motored cabin plane built for heavy, continuous usage under all weather conditions." This was a strong endorsement from an operator in the field.

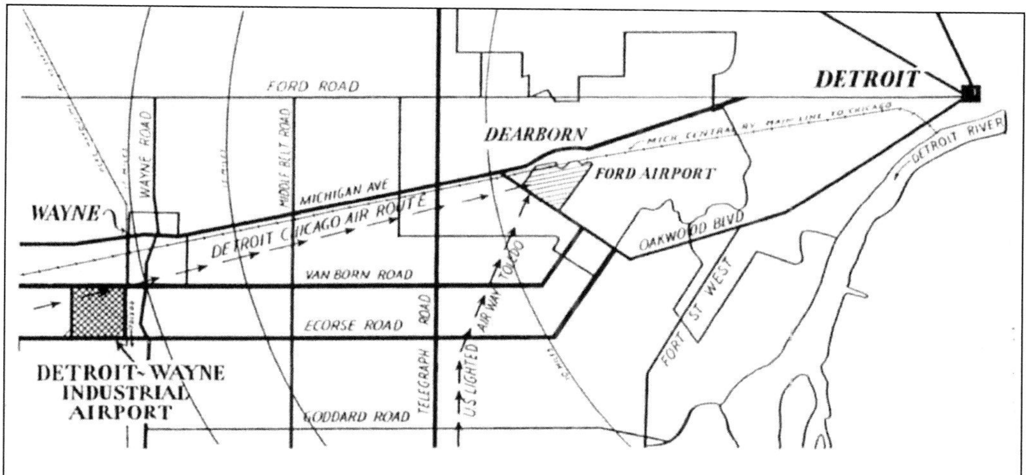

Map Showing Location of Detroit-Wayne Industrial Airport

With 1928 sales increasing, it was clear that the Northville factory could not handle Stinson's production needs. In September, Stinson announced a reorganization and the sale of 140,000 new shares of stock to build a new factory. At the same time, many Stinson shareholders were developing a new airport near Wayne, Michigan (see page 29), designed to fit the manufacturing and terminal needs unmet due to the lack of a Detroit city airport. In December, Stinson confirmed it was building an 85,000 sq. ft. factory at the new Detroit-Wayne Industrial Airport.

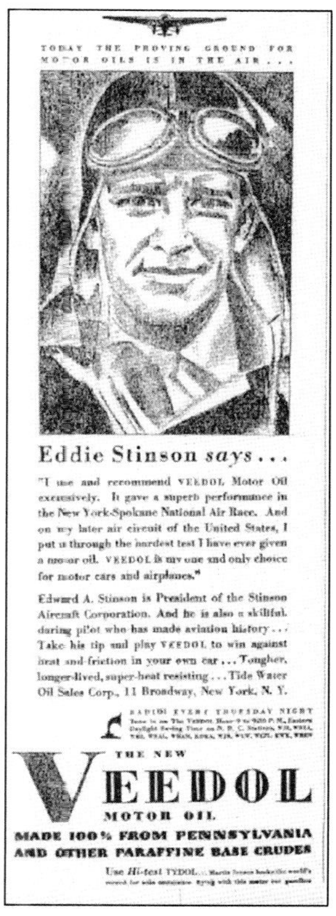

By 1929 Eddie Stinson was a well-known aviation celebrity, well known enough that Veedol Motor Oil ran one column ads in *The New York Times*. (The *Times* is a very expensive newspaper, so the Veedol people must have been sure of Eddie's recognition to use him in an ad.) Whether he got paid in cash or oil is unknown. His notoriety was such that he was also placed in a national testimonial ad for Camel Cigarettes, which was supposed to be his favorite brand.

Eddie Stinson was by all reports a hail fellow well met, liked by most, and apparently unable to say no. Here he has lent his name and likeness, although he looks more like the senior George Bush, to a bubblegum card. It is interesting that the plane he is supposed to be sitting in is armed with two machine guns when, in fact, he never went overseas nor saw aerial combat of any kind. It is unknown how valuable a trading card an Eddie Stinson would have been in those days.

Not to be outdone, Bill Mara lent his visage to the Gordon Aerotogs Company of St. Paul, Minnesota. According to the ad in *Aviation* magazine, Bill Mara had written the company saying, "Your wonderful suede jacket has kept me comfortable on many gusty trips," adding, for the headline, "Remarkable Service and Comfort." The fact that Mara was approached to do the ad indicates his favorable recognition in the aviation industry.

Five

E.L. CORD RESHAPES STINSON

E.L. Cord was, like Henry Ford, a man with a clear idea of the products he wanted to build. Cord products were, however, the direct opposite of the market Ford sought. His automobiles were low volume, stressed style and luxury attributes, and were expensive. In 1930, Cord's Grand Dame, the Duesenberg, cost $9,500 for the chassis fenders, running boards, hood, and dash. The body was not included; the customer was expected to select his or her own custom body style. The cars Cord produced under the Auburn, Cord, and Duesenberg nameplates are today the Queens of the Concours d' Elegance. Cord had bought a Stinson SM-1, learned to fly it, had met Stinson and liked him, and had plans to enter into the world of aviation. Stinson products fit his mold.

In the early part of 1928, Eddie Stinson, shown here with his battered Essex, and his management team came to the realization that their Northville facilities had taken them as far as they could go. If they were to grow, they had to make changes. They were fully aware of the new airfield being developed just a few miles south in Wayne, Michigan. Stinson had some tough choices to make. A move to Wayne meant a major investment in land and buildings, which they had been able to avoid in Northville. And, a move meant turning their backs to Northville, a town that had welcomed them with open arms. It meant leaving people who had gathered together to personally pay for land for the Stinson flying field and had provided many willing workers. It was not an easy choice, but in the end the decision was made to invest in a purpose-built factory at the new Detroit-Wayne Industrial Airport.

The strong-selling Stinson Junior SM-2 soon generated a long list of model designations that were mostly reflections of Approved Type Certificate engine options. Strong sales kept the pressure on the Northville factory to meet demand. Meanwhile, talks were quietly under way with Stinson, the Detroit-Wayne Industrial Airport people, and local banking interests. These meetings were easily arranged since members of the Stinson board of directors were in banking and were also represented on the Detroit-Wayne Industrial Airport board. Finally, in September of 1928 Stinson announced that it had a purpose-built airplane factory under construction on the grounds of the new Detroit-Wayne airport.

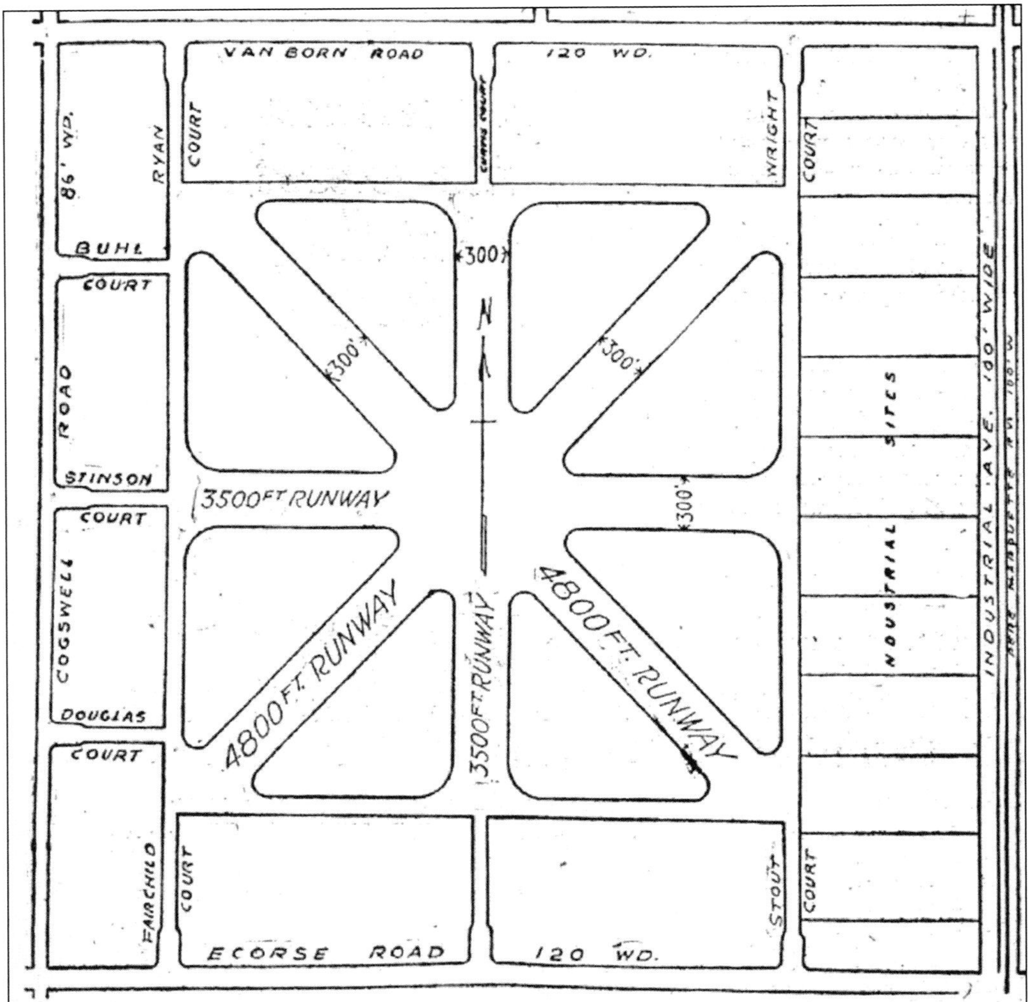

ABOVE: Map of Detroit area showing location of Detroit-Wayne Industrial Airport—a private enterprise backed by Detroit bankers and airmen to provide a co-operative airport for plane builders and operators. Below: Runway layout, covering 280 acres with 360 acres surrounding for factories, shops, hangars and general port facilities.

The map of the Detroit-Wayne Industrial Airport (DWIA) shows what prompted the Stinson organization to consider a move from Northville. Along the DWIA's immediate eastern border was the Pere Marquette railroad track (an important factor in those days), and there was plenty of land for buildings and expansion. With the runways at hand, a completed airplane could be pushed out of the factory right onto the field for testing. Also, the interurban train line that served Northville also served Wayne, meaning that trained workers could get to the new factory and stay with Stinson.

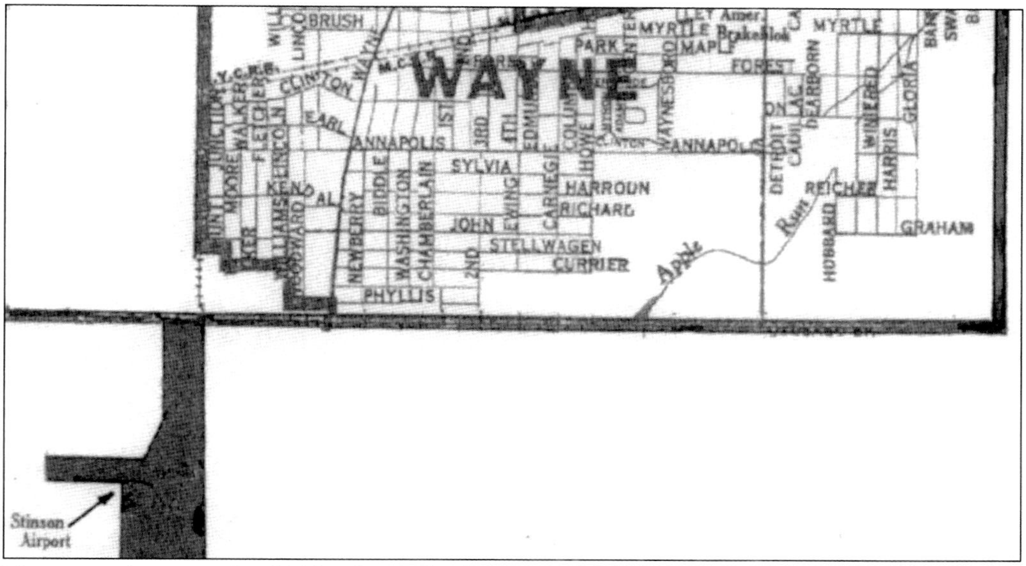

This map, drawn in the 1940s, shows the location of the Stinson Airport as it existed at that time, although the original plan was never completed. The Stinson factory location was on the east side of the field, roughly in the area with the words "Industrial Sites" and opposite the area named "Stinson Court." The deal with the Detroit-Wayne Industrial Airport must have been finalized in 1928.

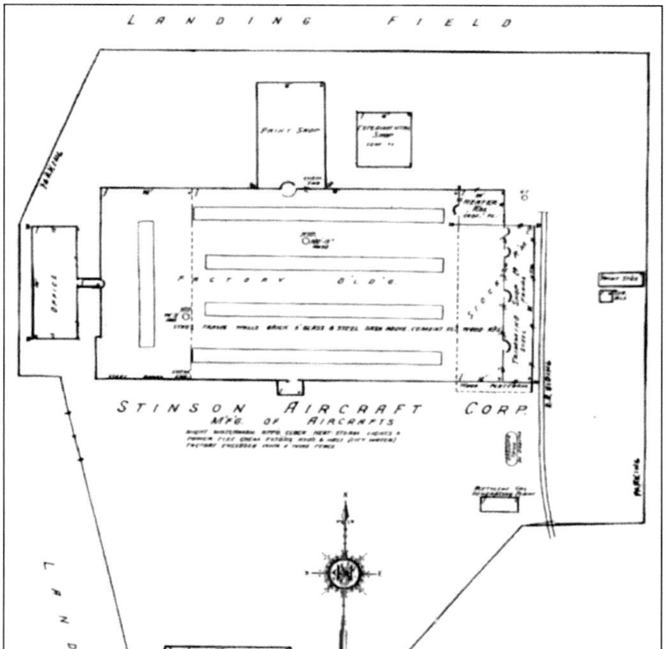

This is the detail drawing of the Stinson Wayne factory as it appeared in a 1930s commercial atlas. The oddly-shaped line around the facility indicates a six-foot-high chain link fence. The building at the bottom of the plot is listed as a service building. Newspaper reports in early December of 1928 indicate factory construction was under way, and there were hopes that it would be completed by January of 1929.

This view of the completed Stinson Wayne plant is looking toward the southeast. The railroad track can be clearly seen at the top of the picture, as can the spur line running to the main factory building. At the far right is the airplane service area. The office building, separated from the main office building by a covered walkway, is at the lower right. Protruding out from the main factory at the lower left is the semi-separate paint department; to the left of that is the experimental shop. the chain link fence, with a wide gate for taxiing airplanes, clearly defines the property.

With this facility in place, it must have been clear to E. L. Cord that Stinson, well-positioned with its strong sales, could not only produce profits but also could be an outlet for E. L. Cord's Lycoming engines and could also produce the kind of airplanes Cord wanted to supply his future airlines ventures. Cord moved swiftly in the summer of 1929 to acquire Stinson. They had thus far escaped the airplane merger mania then rampant.

The news of the Cord-Stinson merger was made public in November of 1929. It was a time of frenetic economic activity; the stock market had just crashed, following a then-unprecedented period of rise, airplane stocks were the dot-coms of the day and mergers and acquisitions were rampant. Stinson Corporation's 123,905 outstanding shares of stock had fallen from a high of $20 in February to a low of $16 on November 19. By comparison, the Detroit Aircraft Corporation fell from $17 on August 23 to $7 on November 29. Eddie Stinson, much liked by E.L. Cord, remained as President, and Bill Mara as Vice President, with E. L. Cord Chairman of the Board.

STINSON MERGES WITH CORD AT CHICAGO MEET

Powerful Corporation Joins With Wayne Concern

The following notice of the merger of the Stinson Aircraft Corporation with the Cord Corporation was received from Chicago yesterday:

Merger of the Stinson Aircraft Corp. with the Cord Corp. has been completed, William A. Mara, vice-president of the Stinson company, announced here Wednesday.

Stinson company will retain its name and personnel as a unit of the Cord interests under the terms of the merger, Mara announced.

E. L. Cord, of the Cord Corp., has been elected chairman of the Stinson board and Edward L. Stinson, veteran flier, president of the Stinson company.

Other members of the Stinson board will include Mara, vice-president; Raymond S. Truitt, secretary, and William Deshield, treasurer.

"Addition of the Stinson company adds an aircraft manufacturing service to the other enterprises of the Cord corporation," Mara said. Other companies of the Cord Corp. include the Auburn Automobile Manufacturing Co. making the Cord front-wheel drive and Auburn cars, the Lycoming Manufacturing Co., making automobile and aviation engines, the Columbia Axle Co., Duesenberg, Inc., and the Limousine Body Co.

Despite the beginning of difficult economic times, Eddie and Estelle Stinson did quite well as a result of the Cord-Stinson merger. According to published reports, Stinson shareholders were offered one share of Cord stock, or $17 cash, for each share of Stinson stock. It is unknown what the original shares had cost, but in a Bill Mara note, Mara mentioned Stinson shareholders getting a 17 to 1 return. Unknown is the amount of shares held by Stinson and Mara. However, they both soon moved into large houses in Dearborn, near the new Wayne plant.

E. L. Cord had many balls in the air during 1929, not the least of which was the launch of his radical front-wheel drive namesake automobile, the L-29 Cord. The car was powered by one of Cord's Lycoming, straight-eight engines delivering 125 horsepower. The stunningly beautiful car had a top price of $3,295; today an "excellent" L-29 will command $175,000 or more. The Cord exemplified style and luxury combined with solid engineering and craftsmanship.

Cord's mighty Duesenberg was another example of his taste in vehicles. Engineered by the Duesenberg race car builders, it was extremely fast in spite of its awesome size. Its straight-eight engine, built by Lycoming, was four feet long and could push the car to 90 miles per hour in second gear. It was the car of choice for movie and sports stars. The Cord, pictured at top, and the Duesenberg both had aircraft overtones in the styling of their instrument panels. Today it is common the see Duesenbergs sell for over $1,000,000.

In June of 1929, E. L. Cord formed the Cord Corporation, a holding company, that gathered all his automotive interests. In July, the Cord Corp. also formed the Corman Corp. to produce the Corman 3000 airplane shown above. The Corman prototype had been developed in 1928, before and without any Stinson involvement. At that same time, the Lycoming Engine Co., a supplier of quality auto engines and ready for the production of radial aircraft engines, became a part of Cord Corporation. With the acquisition of Stinson, Cord put in place his idea of an aviation empire vertically integrating Lycoming engines in Stinson-built airplanes for use in his soon to be started Century Air Lines.

Eddie Stinson is almost dwarfed by his massive SM-6B. Conceived in 1929, it is shown here in front of the new Stinson Wayne factory. This airplane was the biggest, with a wingspan of 52 feet 8 inches; the longest, at 34 feet 4 inches; the heaviest, at 5,350 pounds; the most powerful, with a 440 horsepower engine; and the most expensive single engine airplane, at $19,500, that Stinson ever built. The airplane could carry eight passengers, or if used as a cargo carrier, could lift nearly 2,000 pounds. Note the entry steps, always large on Stinson aircraft, are huge. The SM-6B could cruise at 128 miles per hour, and with a ceiling of 18,000 feet, could clear any mountain in the continental U.S. This airplane was meant to be a workhorse. Only 11 were built by Stinson. A more powerful and stronger version was built in and served in Peru.

10-Place Monoplane

The Stinson Ten-Place Tri-Motor Airliner

Just weeks before the announcement, the stock market crashed and the economy began a slide that would take years to recoup. E.L Cord took it in stride and moved ahead with his plans to be a factor in aviation. Cord had actually gotten into aviation the year before the Stinson acquisition, when he and longtime friend and associate Lou Manning formed the Corman Company to develop a tri-motor airplane using engines being developed by his Lycoming Division. Revealed to the press and shown as a Stinson at the Detroit All American Aircraft Show in the spring of 1930, his Corman tri-motor (renamed the SM-600C, or Stinson Model T) was the first new product of the merger. Using his own Lycoming engines, Cord announced a price of $23,900, a bargain for a nine-passenger airplane. Ford's 5-AT 14-passenger Tri-Motor cost $55,000. However, the annual listing of all commercial airplanes in the August 1930 issue of *Aviation* magazine did not have a SM-6000 listing. Apparently, the airplane was not quite ready for sale at the time it was shown.

The SM-2 Junior went through a bewildering number of designations that mostly reflected alternate engine installations and the requirement to have the variations recognized with a Federal "Approved Type Certificate." The basic dimensions of the airplane were unchanged, with the exception of the massive model SM-6B seen on page 82. Production of models SM-2 through SM-7B was about 150 units.

On January 18, 1930, with the ink barely dry on the Stinson-Cord legal agreements, a small news release in *Air Transportation* magazine told of Bill Mara's order of 500 Lycoming engines for delivery by May 1, 1930. That news was the first manifestation of the vision E.L. Cord had of the integration of his aviation components. The outcome of that news release would be the sensational Stinson SM-8A priced at $5,775, about one-half of its SM-7A predecessor.

The SM-8A hit the softening airplane market like a bombshell. In April, ads that carried the line "Stinson Aircraft Corporation, Division Cord Corporation" for the first time, stressed the $5,775 price tag. The impact of the price was doubly strong because of the combined reputation of Stinson and Cord. Importantly, for the consumer, this was not some tired outdated model being dumped; this was a new Stinson airplane at half the former models' price. SR-8A competitors' comparable models were at a severe disadvantage, with Buhl priced at $11,000; Ryan at $12,985; Cessna $11,000; and Fairchild at $12,900. By the end of April, 32 SR-8As had been sold, and the Stinson brand, supported by Cord, had created a whole new price/value position in the aviation marketplace. It is interesting to note that the two auto companies most deeply involved in aviation were Ford, in Dearborn, and Cord, in Wayne, each with its own airport and residing just a few miles apart.

Stinson kept the marketing pressure on with more ads in June, a month that saw the sales of 75 more SR-8As. The June ads featured the $5,775 price, but additionally blurred the distinction between Lycoming and Stinson and stressed the vast manufacturing capability behind the price. Perhaps because of Cord's automobile experiences and the realization of the importance of qualified dealers the ad also is an offer of some open franchises for "Right Men." Several years previously, Mara had put together a retail package of installment buying and insurance coverage. With sales booming, it seemed that the investment in new facilities and gaining Cord's purchasing power was the correct thing to do. Bill Mara Jr. vaguely remembers the family being photographed with some luggage. It could well be the young Mara family being used as models in these ads.

By the end of 1930, Stinson was outselling all other cabin airplanes in the U.S. Still looking for additional niches in the aviation market, a prototype twin-engine amphibian craft was built. It was somewhat similar to Sikorsky's moderately successful S-38 airplane. Stinson's plane, designated the SM-9, was tested during the summer of 1930, probably at Belleville Lake, a man-made body of water, about two miles south of the Wayne assembly factory. Photos of this airplane are few.

This view of the SM-9 shows the wheels in their awkward retracted position, up out of the water but enclosed in the wing or the fuselage/hull. The object seen on top of the cabin below the booms is probably an open-hatch door. This view also shows the tail wheel, which was probably steerable and acted as a rudder in the water. The plane suffered a crash landing and the project was abandoned. Perhaps the success of the SM-8 was fully occupying the Stinson engineering staff to the exclusion of an "iffy" project.

Tools and equipment from Northville arrived at the Wayne plant Monday, April 1, 1929. According to the Wayne newspaper, it was installed immediately and the first airplane was completed by Wednesday. It seems most likely that completion of the plane was well under way before it arrived in the new factory. That same report also said employment would reach about 300 by the end of the year. This picture, taken June 6, shows the six-foot-high chain-link fence had not yet been installed.

Another photo taken June 6th of the west side of the factory with a finished plane coming out. Ready for testing, the runway was only a few yards away, a vast improvement over the Northville situation. Stinson and Mara were in a great hurry to get the factory in operation in early April. The annual Detroit All American Aircraft Exposition started on the eighth of the month; and they wanted to bring prospective airplane buyers, airline officials, and potential dealers to the brand-new factory, which was probably the best in the nation.

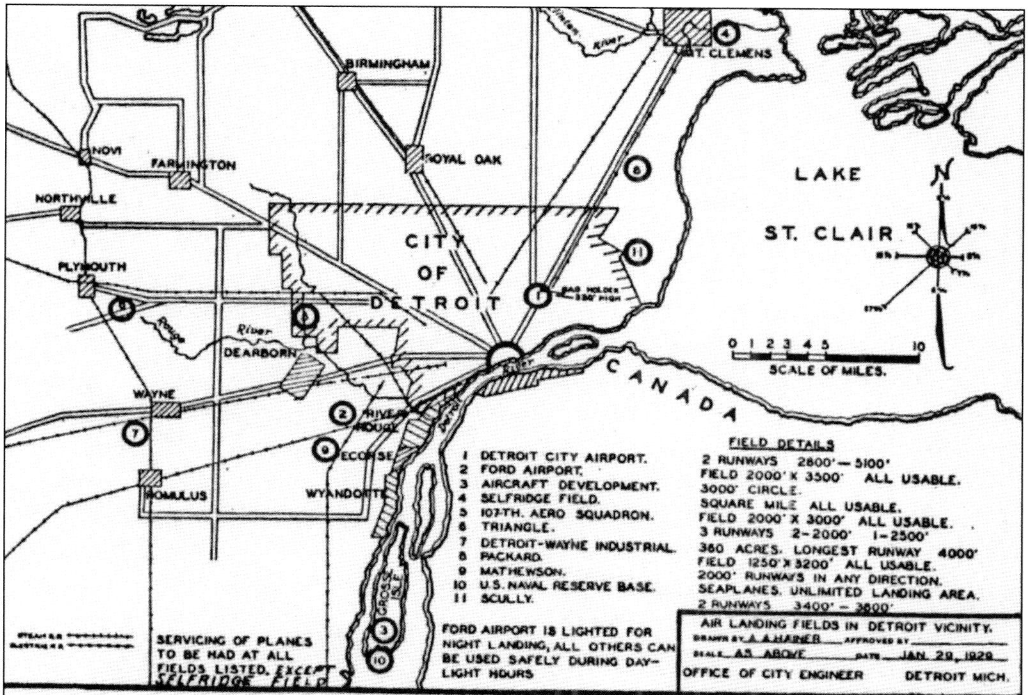

Location of Detroit airports available for visitors flying to show.

Detroit's second All-American Aircraft Exposition attracted airplane manufacturers from all over the nation. The above map was provided to all exhibitors. It is interesting that the former Stinson field at Northville was not shown. The airplane exhibitors displayed 100 airplanes. Stinson showed four: a SM-1; two SM-2 Juniors; and the SM-6A. The show drew big crowds, including 54 men from Wayne who wanted to see what this airplane business was all about. To Stinson and Mara, the show was of great importance, not only as a showcase of their airplanes, but also of the new factory. Additionally, the show gathered Stinson's 45 dealers for the company's first sales meeting. Mara and Sales Manager J.D. Campbell laid out plans for production of six planes a day and the factory marketing support of the production. It was a pressure-packed week for the Stinson organization.

These two men, E.L. Cord and Lou Manning, were the founders of the Corman Corporation, the name an amalgam of their names. Launched in 1928, they intended to build an experimental tri-motor airplane to use the radial air-cooled engines being developed by Cord's Lycoming Division. Lou Manning, then president of Cord's Auburn Motor Car Company, was a close friend of Cord's and an avid pilot who had prodded Cord to learn to fly. The Corman Corporation was incorporated in Dayton, Ohio, and aircraft engineer Horace E. Weihmiller was employed to design and build an eight-passenger transport-type airplane. That airplane, not yet certified by the Federal Government at the time of Cord's Stinson acquisition, received Approved Type Certificate #335 on July 10, 1930, just in time to cinch a fleet sale to the new "Ludington Line" airline. It was the vision of these two men to produce a good, low-cost, multi-engine airplane that launched many budding airlines in the down-market 1930s.

Opposite: The SM-6000B, NC-11153, seen here, was built in 1931 and is owned by Greg Herrick's Yellowstone Aviation Inc. The SM-6000B was a further refinement of the earlier Tri-Motors. The improvements were mostly in increased interior options, including "Club" options for executive-use airplanes. E.L. Cord, wanting deeper aviation integration, started his own airline, Century Airlines, with service between St. Louis and Chicago and Detroit and Cleveland, using 14 of his own SM-6000B airplanes powered with his own Lycoming engines. The price of the SM-6000B was increased to $25,000, which was $1,100 more than the previous model. About 42 B models were sold. Things were happening fast at Stinson; within a few months it had gone from a small-time, one-at-a-time airplane maker to a volume producer of personal planes and a leader in the production of multi-engine airliners.

When Cord acquired Stinson, he brought not only the Lycoming Engine Division; he also came with his own Corman Tri-Motor airplane (see page 76) and the desire to produce it. Renamed, and with some "Stinson" refinements added, the very attractively priced SM-6000 sold well. Of the ten or eleven built, eight of the airplanes were sold to the New York, Philadelphia, and Washington Airway Corporation, better known as the Ludington Line. The planes were used to provide hourly flights between New York and Washington between 8 a.m. and 5 p.m. That schedule competed directly with the railroads and was the beginning of commuter flights. Stinson was now in multi-engine airliner production.

A look up the aisle of Greg Herrick's SM-6000B configured with eight luxury seats; the plane was originally designed to carry from 6 to 10 passengers according to the buyer's wishes. With just two engines running the fully loaded plane, its pilot could maintain an altitude of 6,000 feet. Of great importance in those days of "pasture" airports was the plane's ability to take off with a run of about 700 feet and land at 60 to 65 miles per hour, with a run of about 400 feet.

The passenger compartment has an open airy feeling, the result of large shatterproof windows that could be opened during flight. Additionally, the cabin was equipped with lights. Cabin heaters were optional as was a lavatory room, dual control wheels, radios, night-flying equipment, and wheel pants or fenders. Engine cowl rings were standard as were ground adjustable propellers and electric starters. Goodyear "airwheels" were optional.

The first "all Stinson" tri-motor was the Model U that got its Federal approval May 12, 1932. While superficially similar to previous aircraft, the Model U was completely new. The wingspan had been increased by over 6 feet, the power of each engine increased from 215 to 240 horsepower, and the cruising speed increased from 115 to 123 miles per hour. To obtain more lift, the outboard engines were installed on stub wings and wide "lift" struts supported the wings. Perhaps most importantly, the price of the airplane was $22,900, which was $3,000 less than its predecessors.

Pictured here is another Model U with shorter engine cowl rings. This was an early stage of the usage of the National Advisory Committee for Aeronautics (NACA) research on the beneficial effects of cowlings on radial engine cooling and drag reduction. As can be seen, the whole airplane has been "streamlined," which had a direct effect on speed and range. The forward-slanting windshield, a trend at the time, was intended to prevent instrument panel lights from reflecting on the windshield. The very fact that it was needed indicates that night flying was becoming more prevalent and that airlines were providing "round-the-clock" service.

This is the cockpit of the Stinson Model U. As can be seen, the airplane was equipped with dual controls; however, the co-pilot was soon dispensed with, and that space used the bulky radio equipment of the day. The panel instruments were indirectly lighted to assist in eliminating nighttime reflections. The outboard engine instruments were in the panel instead of outside on the engine nacelles, as was the Ford Tri-Motor style. All control system pulleys were ball bearing type to ease pilot control workloads. Today, most civilian training planes have instrument panels more complex than this ten-passenger airliner. Note the triangular cloth panel at the top of the photo. Equipped with snaps, it could be put into position to keep the sun from beating in through the overhead windows, a simple solution to a simple problem.

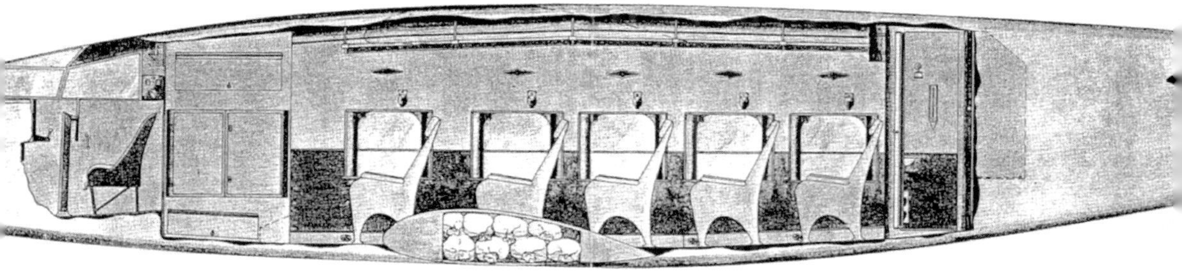

This cutaway drawing shows the standard Model U 10-passenger seating configuration. Above each seat is a reading lamp and above that is a hat rack, an important feature at a time when everyone wore hats. Below the seats are footrests and heating/ventilation vents. At the far right is the lavatory/dressing room. The stub wings storage area is shown filled with mail bags. The compartment between the main cabin and the pilots is intended for passenger baggage.

This definitely non-standard interior is a U series Executive model as ordered by George Hearst, publisher of the *San Francisco Herald*. The plane had, in addition to the main cabin, a lunch room, seen here set for lunch, and a dressing room/toilet. Each main cabin seat had its own radio jack; over the door to the pilot's compartment were an altimeter, air speed indicator, and clock. The cabin design and construction was a collaboration of Stinson and Duesenberg teams. Not surprisingly, the Auburn Automobile Company, a Cord Corp. entity, also had an Executive Model flying office.

The massive Wayne plant tri-motor welding jig is a world apart from its counterpart, produced in Northville just a few years earlier (see page 54). That fixture typifies the rapid development of the Stinson organization. The hands-on-everything company Eddie Stinson and Bill Mara had created in 1926 became a significant division in a major corporation and a leading producer of single and multi-engine airplanes by 1931. Stinson and Mara were quite suddenly custodians of dozens of Stinson dealers, hundreds of workers, a factory and airport, the Stinson brand identity, and the good name of the Cord Corporation.

This sub-assembly welding jig, with parts held rigid for consistency, is in stark contrast to the operation shown on page 53. Clearly, Stinson's Wayne operation was designed and geared for quality, volume production. Selling their products, however good, into a declining market economy was the pervasive problem that Stinson and all other airplane builders faced in the early 1930s.

The mighty Model U towers above the well-dressed poser and puts the size of the airplane in perspective. Born at the beginning of the nation's worst economic depression, 40 Model Us were sold. Originally priced at $22,900, by the end of 1932 the price was $19,500. Before the market crash in 1929, 5,357 commercial airplanes were sold; in 1930, 1,937 were sold; and in 1931, 1,658 were sold. In the two years after the crash, 70 percent of the commercial market was lost, which in turn wiped out many airplane builders. The number of manufacturers went from 92 in 1929 to 46 in 1931; shortly, even the mighty Ford Tri-Motor operation would end. One bright spot was the sale in 1931 of 875 military airplanes. However, Stinson had made no real effort to build "military" airplanes and was counting heavily on the new "R" series of commercial airplanes to survive.

This photo, taken on the morning of January 25, 1932, was the last photo ever taken of Eddie Stinson. It was the day that the ATC for the Model R-2 seen behind Eddie Stinson had arrived. Stinson was still 37 years old on that date, but he looks tired and 10 years older. Perhaps the photo reflects the accumulation of too many cigarettes and too much gin. There are indications that, perhaps, earlier in his life he had contracted TB, had had a nervous breakdown, and was suffering a debilitating chronic illness. One of his Detroit newspaper reporter friends wrote of visiting him in a hospital, after surgery, a few months before this date. Nonetheless, this was the moment of departure when Eddie Stinson, in a pressed suit wearing pearl-grey spats, was ready to set out to demonstrate the R across the country. The first major stop was Chicago.

WEATHER

The Detroit Free Press

1831 — OVER A CENTURY OF SERVICE — 1932

METROPOLITAN FINAL EDITION

Year. No. 267 — TUESDAY, JANUARY 26, 1932 — 24 Pages — THREE CENTS

EDDIE STINSON KILLED

GAS TAX ED TO PAY ERT BONDS

Is Silent on of Group Named

RESS CUTS ERAL COSTS

Kettering Says Boom Awaits New Products

Flow of Hoarded Savings in Response to Awakened Wants Holds Revival Key

By WILLIAM C. RICHARDS

BUYING STRIDE GAINING SPEED AT AUTO SHOW

Dealers Report Sales at Conference on Trade Revival

PROGRESS NOTED IN NEW MODELS

By FRANK D. WEBB

MAHLER TRIAL JURY SPEEDED

Opening Arguments to Start Tuesday

ADMITS DRY LAW FOES' SINCERITY

Dr. Cherrington Here, Prof. s Tolerant

AVIATION LEADER LOSES LIFE

EDDIE STINSON

WALKS CHALK IN HIGH COURT

Heir Asks U. S. Justices for $2,000,000 for Staying Sober Five Years

NEW TRIAL DENIED

FRIENDS VOICE FLIER'S PRAISE

Call Him Industry's Leading Figure

AIR CRASH FATAL TO NOTED DETROIT AVIATION PIONEER

Wing of Plane Sheared Off as He Tries to Land in Chicago Park after His Fuel Is Exhausted

Three Passengers Are Injured when Dean of American Pilots Plunges to His Death; Famed for Numerous Records

Chicago, Jan. 26—(Tuesday)—(A. P.)—Edward A. Stinson, 38, pioneer American airman, died this morning of injuries received when a plane he was demonstrating crashed in Jackson Park last night.

Stinson, known as "The Dean" by reason of having more hours to his credit than any other aviator, was returning to the Wayne, Mich., plant of the Stinson Aircraft Corp., of which he was president, when his ship developed trouble over the Lake Michigan waterfront. In making a landing the plane struck a flagpole and was wrecked, a wing of the craft being sheared off.

Stinson, pinned beneath the wreckage, was extricated and rushed to the Illinois Central Hospital where he died hours later. Three companions, Clark Field, of Kalamazoo, Mich.; John Tompkins, of East Chicago, and Fred Gillies, of Chicago, were only slightly injured.

Stinson Feats Awed World for 20 Years

HELD GENERAL'S RANK

The world woke up Tuesday, the 26th of January, 1932 to learn that one of its famous flyers had been killed. Stinson had taken off with 55 gallons of fuel heading for Chicago to begin a demonstration tour of the R-2 Model. Burning about 12.4 gallons of fuel per hour, he had stopped briefly in South Bend, Indiana. In Chicago, Stinson made a series of demonstration flights; then, near dusk, which occurs early that time of year, he set out on the last short flight of the day. With Stinson were F.M Gilles, Stinson Sales Manager, Clark Field, and J. Tomkins. Shortly after take-off, just over Lake Michigan, the engine lost power. Stinson set up a controlled glide for a landing on the Jackson Park Golf Course and was close to the ground when the right wing hit a flagpole, unseen in the gathering dark. The plane crashed nose down and, being out of gas, did not burn. All four walked away from the wreck. Eddie, more injured than even he knew, died that night in the hospital, his chest crushed. Bill Mara Jr., remembers being at the movies with his family when his father was summoned for an urgent message.

The loss of Eddie Stinson in the R-2 was a devastating personal loss to the Stinson organization, including E.L. Cord, who was "extremely fond" of Eddie Stinson. Nevertheless, life had to go on and there was a new airplane to sell in a tough market. The "R" was essentially a major rework of the good, but tired, Junior design. The wingspan was increased slightly over 1 foot, but the fuselage was shortened by about 5 feet and deepened, and the landing gear was reconfigured. Engine power was the same as earlier models. It was a face-lift that seemed to justify a $600 increase over the cost of the Model S.

The R Model airplanes, from the outset, were planned to offer buyers a choice of fixed or retractable landing gears. The retractable landing gear Model R-3 was about 30 pounds heavier than the R-2 and cost about $500 more, but was otherwise virtually identical. With the wheels retracted, the cruising speed was increased by only 8 miles per hour. The marketing of airplanes was very much on everyone's mind, and R Model ads began carrying the tag line "The Aircraft Standard of the World": a strong variation of the Cadillac Motor Car theme line.

Shortly after Eddie Stinson's death, Lou Manning, a close associate and friend of E.L. Cord, was appointed President of Stinson Aircraft. Naming Manning to that job signaled the importance E.L. placed on his airplane operation. Manning's experience in building and marketing cars was soon visible. New models were soon regularly coming out of the Wayne factory's experimental shop. And, new marketing ideas were soon tried.

In May of 1932, Stinson announced a program aimed at helping their dealers sell airplanes. The idea, titled the "Stinson Air Cab Service," was to sell the concept of hiring Stinson air cabs, just like automotive cabs, and then have the pilot/instructor give the passenger flying lessons during the trip. The assumption was that once a businessman found out how easy it was to fly he would repeat the process, get a license, and be predisposed to buying the then-familiar Stinson airplane.

One of the first new airplanes Stinson brought out under Manning's management was designated the Model O. Chasing every possible chance for a sale, the prototype was rushed to completion in the spring of 1933. The Model O was built to meet a perceived need for military training aircraft for export. While not many O Models were sold, ten or eleven were produced for Honduras, Brazil, and China at a time when every sale was crucial.

While conceived as a training plane, the Model O could also carry light armament. Two small-caliber fixed machine guns could be installed in the forward fuselage firing through the propeller, or as seen here, a flexible gun could be mounted on a ring in the rear cockpit. Both cockpits had a full set of flight instruments. The airplane made use of many production items, first to speed development, and second to keep costs down.

The Stinson Co. again rocked the aviation world in May of 1933 when they announced another new airplane at the unheard-of price of $3,995, F.O.B Wayne, Michigan. The Stinson dealers must have been ecstatic, since no one offered a comparable airplane within hundreds of dollars of the new Stinson Reliant. The airplane was a full four place, powered by the dependable 215 horsepower Lycoming engine, and featured a long list of standard equipment. And, in addition, it was a "Stinson" with all the meaning attached to that brand.

With a cruising speed of 115 miles per hour and price that was about $2,000 less than the previous R series, it was no wonder that nearly 90 SRs were sold in 1933, in spite of the deepening economic depression. The 1933 Stinson Reliant, named by Bill Mara, was the first in a series of airplanes that for years would grow in performance, luxury, and stature through constant development. According to the National Stinson Club, nearly 1,500 were built. Today they reign as the Queens of any antique airplane fly-in display.

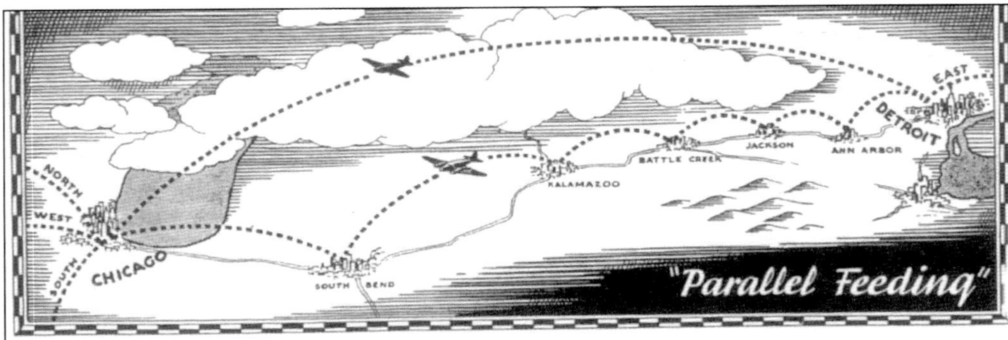

From Kalamazoo Direct to - -

NEW YORK - BOSTON - WASHINGTON - MIAMI
NEW ORLEANS - DALLAS - WEST COAST

STINSON TRIMOTORED AIRLINERS, operated by American Airlines, have reopened a great transportation market in the five prosperous cities between Detroit and Chicago.

A year ago Kalamazoo travelers lacked fast air service. If they wished to visit New York they generally used the railroad. Result, they lost time and the Airline lost revenue.

Now the people of all five cities may leave home or office at the close of the day, are speeded to Detroit on American Airlines' Stinson Trimotored Airliner, make quick connection there with a through plane and arrive in New York at 10:20 P. M.

They may leave home in the morning, spend the day in Chicago and return home in the evening, or make fast connections at Chicago arriving in St. Louis, Louisville, Tulsa or Oklahoma City that afternoon; in Atlanta, New Orleans, Miami, Dallas or Ft. Worth that evening and in Los Angeles in time for work the next morning.

This fine service has been made possible by the Stinson Trimotored Airliner, fastest of its type in America, and built specially to give high-speed service on frequent stop runs and to "Parallel Feed" the through services.

"Parallel Feeding" means providing modern high-speed service to hundreds of important cities on main Airlines which are now neglected or poorly served. These cities are entitled to and will support better service, just as Kalamazoo and its sister cities have responded to the time saving schedules made possible by Stinson Airliners.

To better serve their territories Delta and Central Airlines have ordered fleets of Stinson Trimotors and will shortly inaugurate the fastest services between Dallas, Tyler, Shreveport, Monroe, Jackson, Meridian, Birmingham and Atlanta; and between Detroit, Cleveland, Akron, Pittsburgh and Washington.

THE STINSON TRIMOTORED AIRLINER, AMERICA'S FASTEST AND MOST ECONOMICAL TRIMOTOR

STINSON AIRCRAFT CORPORATION
Division Aviation Manufacturing Corp.
WAYNE • MICHIGAN • U. S. A

Stinson again surprised the industry when it announced a new Tri-Motor, the Model A, in the fall of 1933. The plane was designed to meet a request from American Airways for a low-cost, feeder-line sized airplane with short-field capabilities. The ad describes the plane's purpose perfectly when it says, "it will give high-speed service on frequent stop runs." The Model A brought multi-engine "airliner" service to many small towns that had never before scheduled airline service.

Stinson's new tri-motor was of conventional welded steel tube construction covered with fabric. That made for a rugged, easy-to-build, low-cost airplane. Fully loaded, the Model A could take off in less than 800 feet and land in 400 feet. It could carry eight passengers at a cruising speed of 163 miles per hour, land at 63 miles per hour, and it cost only $37,500. With the new tri-motor, Stinson was back in the airliner business, just as Ford was going out of the airplane business.

Stinson's Model A was designed as a feeder-line transport, meaning many short flights to towns along the line to major cities. The plane was subjected to dozens of take-offs and landings every day. It had to be strong and dependable. This photo of a portion of a Model A shows it being tested with 15,400 pounds of sandbags and not failing. Another fail-safe item was the landing gear that left the wheels semi-exposed to minimize damage in a wheels-up landing. This was a practice followed by Boeing and Douglas.

This is a rare view of the Model A cockpit. Note the improvements over earlier tri-motors, the use of quilted padding in the ceiling and under the windshield, which could be slid open for ventilation or visibility in bad weather. The cockpit of the Model A was well behind the outboard propellers' arc, cutting the noise level for the pilots. With the addition of flaps and retracting landing gears, the control pedestal between the seats fairly bristled with knobs and switches.

The wide-tread landing gear is well displayed here, as are some other features that were soon to be Stinson hallmarks. The engines are closely cowled for drag reduction and better engine cooling, and to get the greatest effect, the protruding engine rocker arms are enclosed in streamlined "blisters." The wing, which tapered in both thickness and planform, was soon given the name "Gullwing." That appellation would be applied to a long series of Stinson airplanes not yet born when the Model A was designed.

The interior of the Model A could have been called a "wide-body" in 1934 with its three-across seating. The unusual shape of the cockpit door was determined by the placement of structural members running up to the wing strut attachment points. As can be seen, passengers were provided with wall-mounted reading lamps, ash trays, ventilation outlets, and sliding window curtains. The above picture shows what appears to be a standard interior.

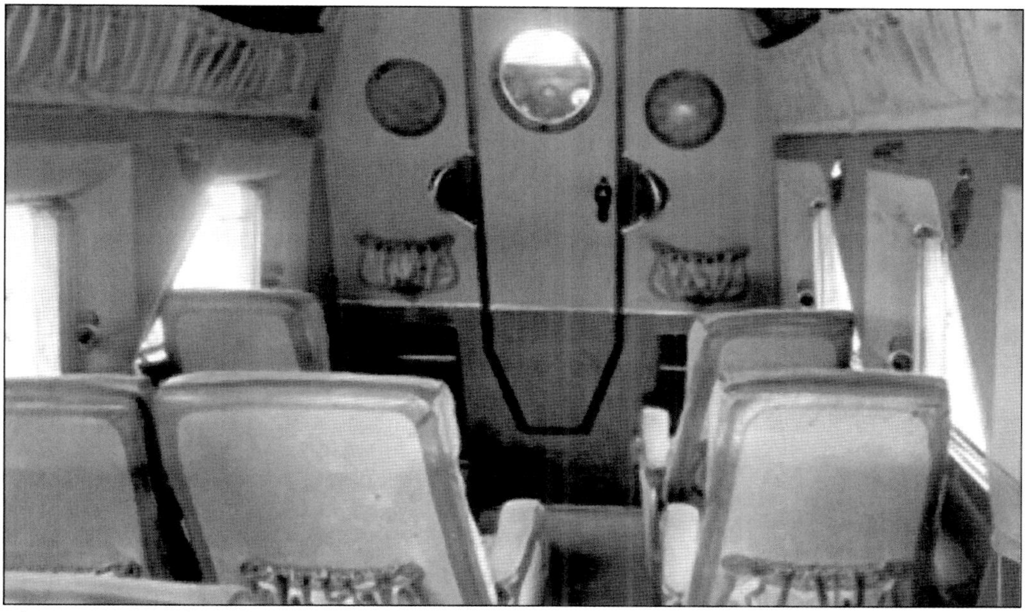

This appears to be a deluxe interior. The seats are bigger, each is equipped with a seat back pouch, or for the front two seats, bulkhead mounted; the hat rack is gathered fabric instead of elastic. There is just a more deluxe feel to this cabin. It should be remembered that passengers were not in the seats very long. For example, Delta Air Lines, the first Model "A" purchaser, used them on the Dallas-Atlanta route, in which people got on and off the plane in Tyler, Shreveport, Monroe, Jackson, Meridian, and Birmingham.

The R.J. Reynolds Corporation bought this custom interior Model A Executive-Transport. This view looks forward toward the pilot's cabin. Special features are the electric fan and the thermos bottle at the right, and a small writing desk on the left. There is a cloak room with a two-way radio behind the bulkhead. The table is removable and the seats can be swiveled to open an aisle. The large plush seats and their ability to swivel limits the passenger capacity to six persons.

A look to the rear of the Reynolds cabin, above the map are what appear to be a compass and a clock. Behind the bulkhead is a full-size sleeping berth and behind that a lavatory. These seats also swivel. About 30 of the Model As were built between 1934 and 1936; they did the job they were designed to do, and they contributed to Stinson's fortunes. However, they were born in a time of rapid aeronautical change and were soon made obsolete by the planes being developed by the Boeing, Douglas, Lockheed, and Vultee gang out west.

Six

STINSON'S RELIANT YEARS

Launched in 1933 with the sensational price of $3,995, the SR series would keep Stinson in the forefront of airplane builders. With the introduction of the SR-5, the following year solidified Stinson's position and provided an airplane platform with the capability to grow and develop. It was the light at the end of a long, dark economic tunnel. Bill Mara was the only management figure still standing after starting the business with Eddie Stinson nine tough years earlier.

The 1934 SR-5 Reliant looked much like its predecessor but it was a considerably different airplane. The wingspan had been shortened by about 2 feet and the chord increased. The Lycoming R-680, 225 horsepower engine had been standardized and wing flaps were available for the first time on a Stinson airplane in the SR-5A model. Also, the price of the SR-5 was $1,780 more than the SR model. Although the basic construction of welded steel tubes with fabric covering remained the same, many refinements added content to the SR-5s. There was more cabin insulation, a better heating and cooling system, and a real attempt to add automobile type colors, fabrics, and design to the cabin. Aerodynamic efficiency was improved by extensive use of fillets where surfaces met the fuselage. The SR-5 series spawned a bewildering number of variants, from 5A through 5F. The differences were mostly in the engine and other options; all required a different Federal Approved Type Certificate.

Opposite: Stinson was so sure of the appeal of their new model that they began a national advertising campaign a month before the official ATC was issued. The main marketing theme advertised the airplane as an "Airliner" for the businessman and the commercial operator. Meanwhile, over the previous few years, E.L. Cord acquired control of AVCO Corporation through convoluted stock trading. Entering 1933, Cord's holdings were Auburn Motor Car, $18.4 million; AVCO, $15.8 million; Cord Corp., $17.6 million; Lycoming, $5.3 million; and Stinson, $730,000. The Stinson organization was a small fish in a carnivore tank, but in 1933, it outsold all 2–6 passenger planes combined.

MARCH, 1934

55

THE ONLY 4-Passenger Cabin Plane Built to "Airline Transport" Standards

THE NEW STINSON RELIANT FOR 1934 IS AN "AIRLINER" FOR THE BUSINESS MAN AND COMMERCIAL OPERATOR

Stinson Leads—

IN SALES

Outstanding value was responsible for the fact that the 1933 Stinson Reliant outsold ALL MAKES of two, four, and six-passenger cabin planes COMBINED, since its introduction in May. Never before was such decided preference shown for any make of plane.

IN SAFETY

With more Stinson Reliants flying in 1933 than any make of four-passenger cabin plane, not a single person was injured and only one Reliant was damaged, in a landing due to heavy fog. This is a tribute to Reliant Flyers and is definite proof of safety.

IN RELIABILITY

By flying more than 60 million miles on single engine and tri-motored planes in the United States, Alaska, South and Central America, Europe, the Philippines and other countries, the Lycoming Transport engine has been proven MORE RELIABLE than any motor of its power class.

The New Stinson Reliant

The new 1934 Stinson Reliant, four-passenger cabin plane, offers many new mechanical features and comforts previously found only in Airline Transport planes, or in planes in the $10,000.00 price class.

The new Reliant provides the utmost in reliable cross-country transportation, with the safety and precision of the modern Airliner, and at operating costs comparable to motor car travel.

The Reliant is the only four-passenger cabin plane built by an organization which also manufactures Tri-motored Transport planes which have proven their reliability in 20,000,000 miles of scheduled Airline service.

Only in the Reliant do you obtain the Lycoming 225 H. P. Transport Motor, whose reliability has been amply proven by 60,000,000 miles of flight in Airline and private service. No other motor of this horse power has ever flown so far. Because safety in an airplane depends particularly on the reliability of the motor, you will be wise to demand that the motor which powers your plane be proven by "Several million miles of regularly scheduled Airline Service."

STINSON Aircraft Corp.
Wayne, Mich., U. S. A.

Division of Cord Corporation

WRITE FOR YOUR COPY OF STINSON PLANE NEWS, MAILED FREE POSTPAID

The 1934 SR-5 came equipped with a wide range of standard equipment. In the auto-like cabin there were ash trays, a glove box, map pockets, carpeting, and roll-down safety glass windows. The control wheels were adjustable and the right wheel removable; a tool kit was supplied and fittings for pontoons were also standard. The basic color was black fuselage with silver or yellow for the wings; other colors were optional.

Cord's Lycoming Engine Division acquired the Smith Propeller Company in 1933. Not surprisingly, Stinson offered the Smith Controllable Pitch Propeller as an option in 1934. The propeller was adjusted by the pilot just as a driver shifts gears, low to start, shifting to high to cruise. For an airplane, the benefit is more load-carrying ability, not an increase in speed. Stinson would install the propeller on all existing Lycoming-engine Stinsons for $975 plus installation, or $775 on new orders.

One of the strong product advantages of the SR-5 was the inclusion of wing flaps as standard equipment. They were a feature just coming into common use on airliners but most private/businessman pilots were unfamiliar with or leery of them. Stinson ran a series of national ads explaining the feature and devoted a full page of their *Stinson Plane News* to the topic. Landing flaps or, as Stinson called them, "speed arresters" were soon to be standard equipment on all but the most basic of training planes.

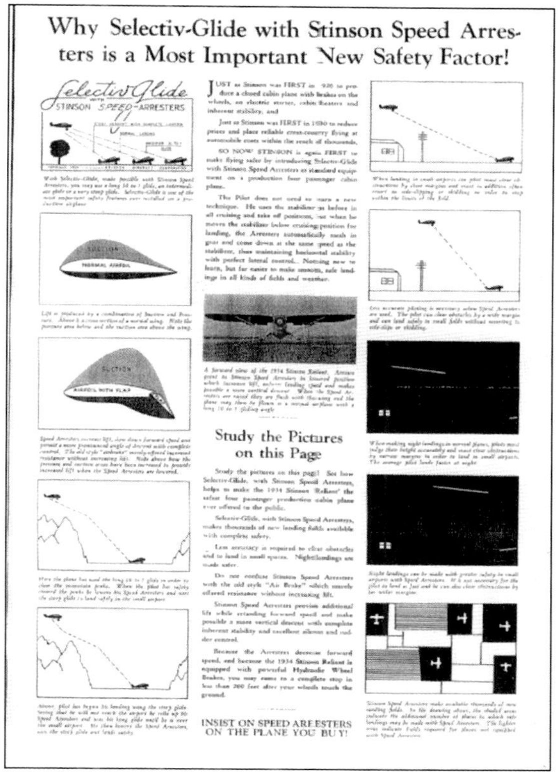

Why Selectiv-Glide with Stinson Speed Arresters is a Most Important New Safety Factor!

Selectiv-Glide with Stinson Speed-Arresters

JUST as Stinson was FIRST in 1926 to produce a closed cabin plane with brakes on the wheels, an electric starter, cabin heaters and inherent stability, and

Just as Stinson was FIRST in 1930 to reduce prices and place reliable cross-country flying at automobile costs within the reach of thousands,

SO NOW STINSON is again FIRST to make flying safer by introducing Selectiv-Glide with Stinson Speed Arresters as standard equipment on a production four passenger cabin plane.

The Pilot does not need to learn a new technique. He uses the stabilizer as before in all cruising and take off positions, but when he moves the stabilizer below cruising position for landing, the Arresters automatically mesh in gear and come down at the same speed as the stabilizer, thus maintaining horizontal stability with perfect lateral control. Nothing new to learn, but far easier to make smooth, safe landings in all kinds of fields and weather.

A forward view of the 1934 Stinson Reliant. Arrows point to Stinson Speed Arresters in lowered position which increase lift, reduce landing speed and makes possible a more vertical descent. When the Speed Arresters are raised they are flush with the wing and the plane may then be flown as a normal airplane with a long 10 to 1 gliding angle.

Study the Pictures on this Page

Study the pictures on this page! See how Selectiv-Glide, with Stinson Speed Arresters, helps to make the 1934 Stinson 'Reliant' the safest four passenger production cabin plane ever offered to the public.

Selectiv-Glide, with Stinson Speed Arresters, makes thousands of new landing fields available with complete safety.

Less accuracy is required to clear obstacles and to land in small spaces. Night landings are made safer.

Do not confuse Stinson Speed Arresters with the old style "Air Brake" which merely offered resistance without increasing lift.

Stinson Speed Arresters provide additional lift while retarding forward speed and make possible a more vertical descent with complete inherent stability and excellent aileron and rudder control.

Because the Arresters decrease forward speed, and because the 1934 Stinson Reliant is equipped with powerful Hydraulic Wheel Brakes, you may come to a complete stop in less than 200 feet after your wheels touch the ground.

INSIST ON SPEED ARRESTERS ON THE PLANE YOU BUY!

Helping Stinson dealers sell airplanes continued to be a marketing goal. To add impact to the Air Cab Service idea, the factory devised a special Air Cab Model painted International Orange and Bonnet Blue with an Air Cab logo on the fuselage. Special payment terms with payments coming out of earnings were available, and Stinson was actively seeking new Air Cab Franchises with operators that measured up to their high standards.

There were extra features all around the SR-5 Models. One of the most unusual was the rear view mirror shown at the left. It was crank-operated and functioned as a vent as well as providing a view to the rear. Another unusual item was the throttle mounted on the control column; on most airplanes they protruded from the instrument panel. The right control wheel was constructed for easy removal, should the owner desire.

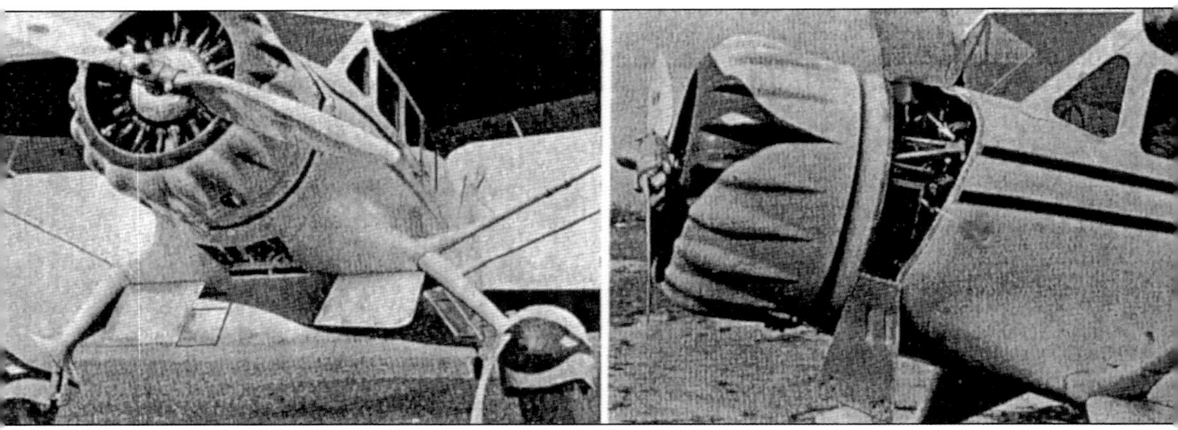

The Model 5 was built with the need for easy inspection and service a priority. Part of the high cost of airplane ownership is that of required inspections. This airplane helped reduce those costs by building in large inspection and service panels. The fabric-covered wings had zipper openings at crucial inspection points. Airplanes without zippered inspection panels often had to have the fabric cut to see vital interior areas.

In 1936 the Stinson SR-7 Reliant "Gullwing" was introduced. It would become the airplane that would be identified by most people as "the" Stinson airplane. The Gullwing came into being through an attempt to keep the product line fresh. Stinson needed an updated Reliant. Rather than producing just cosmetic changes, they opted for a major redesign. The double taper wing, which had been so successful on the Model A Tri-Motor, was rescaled and mated to a SR-6 fuselage and empennage that had been redesigned shortly before. The result was an airplane in the Duesenberg tradition. Seats were leather-covered and available in four colors, window outline trim moldings were stamped and painted woodgrain, and all windows could be cranked up or down. The airplane was finished with 12 coats of high luster paint; there were 5 engines offered, ranging from 225 to 320 horsepower. Even though the SR-7 cost $490 more than the SR-6, it was deemed well worth the money and the airplane sold well.

Other airplanes being developed in 1936 were the Boeing B-17, England's Hurricane and Spitfire, and Germany's ME-109.

Factory advertising supporting the launch of the new crop of SR-7s and SR-8s began in April of 1936 and ran in national aviation magazines. The ad tag line now said: "Stinson Aircraft Corporation, Division of Aviation Manufacturing Corp." That distinction probably did not mean much to Stinson buyers, but it must have had an impact on Bill Mara, who was then working for the fourth organization to control what he and Eddie Stinson started.

Floatplane versions of Stinson products had long been popular. This is a view of Stinson's seaplane operation on Belleville Lake, a man-made lake just three and a half miles south of the Wayne factory. Over the next few years, the Reliant series continually improved, reaching its culmination with the SR-10K in 1941. The Reliant "Gullwings" had cemented a brand image of Stinson power, price, and luxury, but by the late 1930s, Bill Mara was sensing a changing mood in the market.

Seven

Pre-War Boom, Post-War Boom

By 1938, the civilian airplane market was splitting into two segments roughly along the line of engine type. There were the traditional builders of dependable, big, expensive radial engine airplanes and a new group of cheaper airplanes powered by suppliers of small flat-four air-cooled engines. The Stinson Reliant is probably the best example of the former, the Piper Cub the latter. Bill Mara, sensing an opportunity, pulled together M.A. Mills, L.E. Reisner, and Professsor Peter Altman, who had consulted with Stinson before, to design a modern light airplane. The plane had to maintain Stinson attributes, use a new Lycoming 75 horsepower engine, have a low cost, and be easy to build. By May of 1939, the Stinson 105 had its ATC and preproduction models were sent out on demonstration tours. The tour was an outstanding success; priced at $2,995, orders poured in. Perhaps aiding sales was the worsening war situation in Europe, which stimulated vast interest in aviation.

Like other Stinsons before it, the 105 was a lot of airplane for the money. It had wing slots and flaps as standard equipment, which enabled the plane to take off and land in short home fields. The wing slats also improved low-speed handling; the 105 landed at 43 miles per hour, and made the airplane virtually spin proof. Early models lacked power, but soon 80 horsepower engines were available and performance picked up. Seating three, the 105's cruising speed was 100 miles per hour with a range of 350 miles.

The instrument panel has 1940s kitchen table styling but provides the necessary engine/flight information. On the corporate side, Cord left AVCO in 1938; Mr. Victor Emanuel was the new leader. Lycoming, Stinson, and Vultee Aircraft were grouped in the Aviation Manufacturing Corporation. Meanwhile, a war had started in Europe, military orders were arriving, and civilian orders for the 105 were pouring in. The long famine in airplane sales was suddenly over and boom times had begun.

In spite of being a "downsized" Stinson, celebrities flocked to buy the 105. Pictured here is famous driver Wilber Shaw and Bill Mara's brother Francis; Jimmy Stewart and Howard Hughes were also buyers. By 1939, AVCO, concerned that the Wayne plant could not handle the flood of orders for the 105, decided to build a new Stinson factory in Nashville, Tennessee. The plant, despite interrupting the lives of many Wayne, Michigan employees, was operating by spring of 1940.

The rapid acceptance of the model 105 brought smiles to everyone at Stinson. Compared to the SR-10, the new model was smaller in wingspan by 7 feet, and smaller in length by nearly 8 feet, its engine had less power by about 150 horsepower, it was 2,400 pounds lighter, its cruising speed was about 50 miles per hour lower, and the cost was about $9,500 less. Stinson did not skimp in the construction of its lower-priced airplane; it had a welded steel fuselage, solid spruce wing spars with stamped aluminum ribs, and an empennage that needed no drag-producing braces.

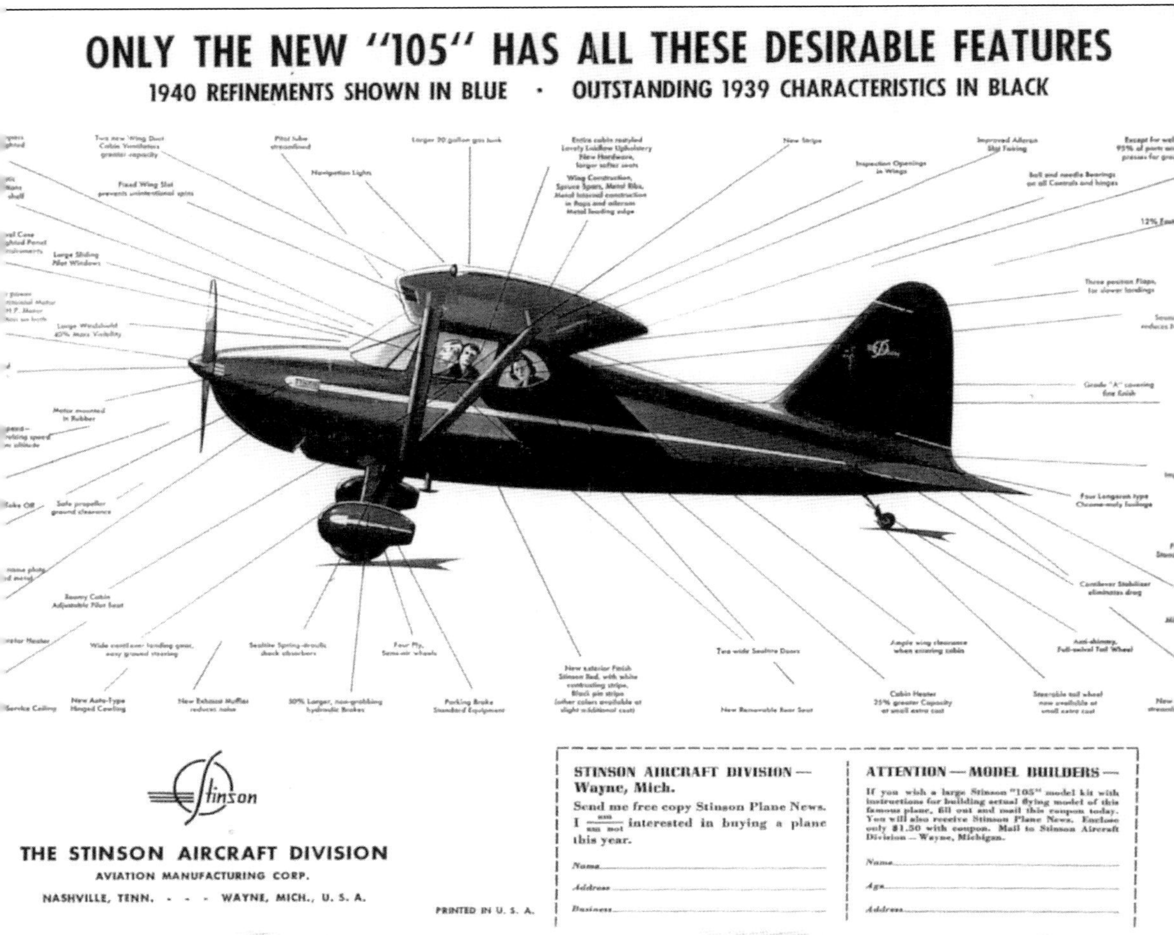

Taking an automotive approach, Stinson provided their dealers with sales promotion materials that detailed in depth the Model 105 advantages, a total of 52 were listed here. Stinson, like most of the car companies, took liberties with the portrayal of people in vehicles; it must have been a family of midgets inside the 105 shown here. Stinson was by then under the control of Vultee Aircraft (note the change in the Stinson logo design, it is in the Vultee style); and as the U.S. got closer to war, military aircraft orders were pouring in on Vultee. Vultee won an order for hundreds of Vengence dive bombers, and Stinson had also gotten an order for 100 of its "Vigilant" Army observation planes. Faced with even more orders on the horizon, Vultee decided to dedicate the new Nashville plant to strictly military airplane production and to return Stinson's 105 program to Wayne. So it was after just a few months away that Stinson returned to Wayne, Michigan with piles of orders, to find suitable workers in short supply.

Hard on the heels of the 105's success, Stinson introduced a newer, better model 10-A that responded to field requests. Named the Voyager by Bill Mara, the 10-A's major improvements were increased power, a shortened nose cowling for better ground visibility, and ball bearing controls. The deluxe version, an automotive concept, offered an electrical system with a generator, electric engine starter, and shielding for radios, an increasingly important item. Over 760 Voyagers were built in the short time between its introduction and the end of civilian production in 1942.

This is an inside view of the Wayne Stinson factory on March 6 of 1941. War was just nine months and one day away. Voyager production was in full sway after the brief exodus to Nashville. This view shows a hospital like-atmosphere when compared to the original dungeon-like factory in Northville. No matter the environment, Stinson always designed excellent airplanes and Stinson workers always built them to that standard.

Responding to a U.S. Army request for an observation airplane with short take-off and landing capabilities, Stinson produced the Model 74 (Army designation L-1 Vigilant) and was rewarded with a contract for over 100 airplanes. The Stinson-designed plane was built in the Vultee Nashville plant along with the Vultee Vengeance military plane. It was a big airplane with a wingspan of nearly 52 feet, and it could take off and land within the confines of a 200 foot circle. It did all that the Army asked for.

Amphibian-type floats added to the L-1's versatility and were particularly useful in the Pacific Theater of Operations. Without floats, the plane was priced at $25,420, making it the most expensive single-engine airplane Stinson ever built. Over 300 L-1s were built. Many of the L-1s were converted into ambulance airplanes. In spite of its huge size and modest 295 horsepower, the L-1 had a top speed of 122 miles per hour.

The Army, wanting a lighter and less expensive liaison airplane, was interested in Stinson's development of the Model 76, a derivative of the civilian Model 10. First flown in the summer of 1941, the Model 76, military designation L-5, performed exceptionally well, and after some modifications to improve its spin recovery, it was put into production. Over 3,600 L-5s were ordered, making it the most widely-used utility airplane in World War II, and spreading the Stinson name to every part of the globe.

The L-5 was such a good airplane for the military that it survived the transition from Army Air Force to U.S. Air Force. The plane also had the distinction of being in front line service in World War II and five years later in the Korean Conflict. The picture above shows the L-5 in Korean markings serving somewhere in Korea. Considering that the original design was completed in 1941, it is remarkable that the L-5 outlasted most of its competitors.

During World War II Stinson came under another corporate leader when Vultee merged with Consolidated Aircraft, becoming Convair, and was headquartered in California. In spite of the corporate shuffling, Stinson was ready for the expected post-war private airplane boom with its Model 108 Voyager. Stinson had 100 distributors and dealers in place to sell and service the planes that were soon to flow from the Stinson factory. Gas rationing was over, the GIs were coming home, and the sky was limitless.

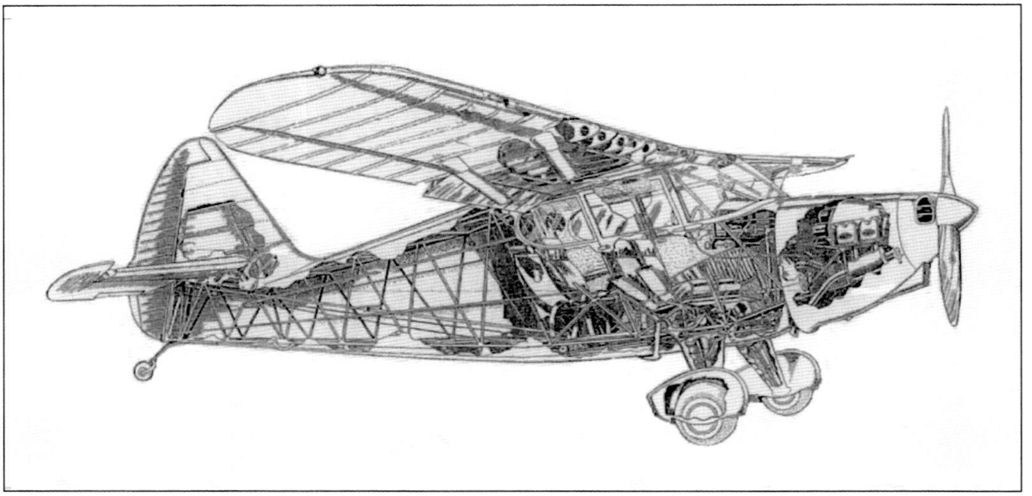

Like all Stinsons before it, the Voyager fuselage was built with welded steel tubing, then fabric-covered; it was strong but heavy. The light plane industry, however, was beginning to move toward all metal monocoque fuselages that were pioneered years earlier and had reached full stride in World War II. They were just as strong but lighter, and once the techniques were learned, they required fewer man-hours to build. The Stinson approach to building airplanes was sound, proven, and inspired confidence, but change was in the air.

This is the way the post-war light plane market was supposed to be. A quick trip cross country to your place in the country, landing on your own lawn near the lake for a weekend of fun. Stinson's Voyager was the airplane immediately after the war to fulfill that dream. The plane cruised at 125 mph, had a range of 500 miles, it could carry four people and luggage, and with the standard Franklin engine it cost $5,495. The Model 108 Voyager was, in fact, the most popular Stinson ever made; 5,500 were built between 1946 and 1948. For a time Stinson was the only profitable division in the Convair conglomerate. Then the entire light plane market collapsed.

Far away in California, Stinson's corporate controller, Convair was having its own million dollar problems with a number of U.S. Air Force programs and was in deep debt. For them, it was expedient to sell small-size Stinson to the Piper Airplane Company to raise cash and simplify their span of control. Piper bought 200 completed Voyagers and assembled about 150 more from existing parts at the Wayne plant. In April of 1949, the Stinson organization ceased to exist. During its lifetime, Stinson built more airplanes for a longer period of time than any other Michigan airplane company. In the end only Bill Mara, who was there at the beginning, was still standing to turn out the lights.

Fifty years after the last one was built, the Stinson name is still respected by people who know airplanes.

Not every promising Stinson design, however, made it to production. Shown above is an artist's conception of the 1936 Stinson Model B Bi-Motor. It was to be a six-seat version of the well-received Model A Tri-Motor. According to B.D. Deweese, the plane, which was priced at $19,950, was intended for private executive use or as a feeder line airplane. It was estimated that the cruising speed would be 155 miles per hour and the range 500 miles. Quite a respectable performance for the day.

This one-off airplane Model M was perhaps built in response to other low wings of similar design being brought forth in 1932. This Stinson has an unusual exhaust system that emptied via short pipes from each cylinder. Being a prototype, it is not clear whether the large gap between the cowl and the fuselage would be covered with adjustable cooling flaps, which was the norm. The heavily "panted" landing gear was an attempt to cut down drag without retracting the gear.

One on the most interesting airplanes never to see production was the Stinson Model 106. It was apparently flown in March of 1946, just six months after the end of World War II. Powered by a Franklin air-cooled engine, it was one of a number of post-war civilian airplanes planned for development. The plane was registered NX 40004, and was apparently test-flown. In the picture above, the front fuselage looks much like the World War II Bell P39, which also had an engine mounted behind the cockpit.

This non-flying mock-up differs from the above in that the forward fuselage seems deeper and more rounded and the windshield seems more upright. Also, there is a definite cooling inlet bulge on top behind the cabin area, the tail booms are at a definite angle and the wheels have "pants." There are no reports of the performance, if any, of the aircraft, but it does look like an airplane that, with the Stinson name, could have created a market for itself.

In a sense, Eddie Stinson never had a conventional job. He had a passion, among many, for flying airplanes. For nearly twenty years, in aviation's most formative time, that's what he did. He had the motor skills and eyesight that would be envied by any athlete, and the immeasurable ability to make the quirky, cranky early airplanes do what he wanted them to do. Although largely unschooled, he had an instinctive ability to conceive airplanes that combined all the needs that he knew were unmet, and the good sense to trust their realization to aeronautical engineers. His fondness for alcohol is reported from so many sources that the truth must lie therein. Undiscovered are any reports of meanness or hostility while afflicted. All describe Eddie as a good fellow to be around; he was at his best demonstrating and selling his namesake airplanes. In truth, Eddie Stinson, as president of a corporation grown large, was probably too often away from his desk, flying and selling.

When he died unexpectedly, millionaire and boss E.L. Cord was reported devastated by the loss of his friend; Bill Mara, his friend and partner for nearly 20 years, was likewise moved. Stinson was also mourned by shop floor workers and the line boys who fueled his airplane. On the day of his funeral, his adopted city, Detroit, gave him homage fit for a head of state. Perhaps the best tribute to the man is that a number of the 13,598 airplanes that bear his name, now more that 50 years old, are still flying.